1,000,000 Books

are available to read at

Forgotten Books

www.ForgottenBooks.com

Read online
Download PDF
Purchase in print

ISBN 978-1-333-00681-5
PIBN 10448978

This book is a reproduction of an important historical work. Forgotten Books uses state-of-the-art technology to digitally reconstruct the work, preserving the original format whilst repairing imperfections present in the aged copy. In rare cases, an imperfection in the original, such as a blemish or missing page, may be replicated in our edition. We do, however, repair the vast majority of imperfections successfully; any imperfections that remain are intentionally left to preserve the state of such historical works.

Forgotten Books is a registered trademark of FB &c Ltd.
Copyright © 2018 FB &c Ltd.
FB &c Ltd, Dalton House, 60 Windsor Avenue, London, SW19 2RR.
Company number 08720141. Registered in England and Wales.

For support please visit www.forgottenbooks.com

1 MONTH OF FREE READING

at
www.ForgottenBooks.com

By purchasing this book you are eligible for one month membership to ForgottenBooks.com, giving you unlimited access to our entire collection of over 1,000,000 titles via our web site and mobile apps.

To claim your free month visit:
www.forgottenbooks.com/free448978

* Offer is valid for 45 days from date of purchase. Terms and conditions apply.

English
Français
Deutsche
Italiano
Español
Português

www.forgottenbooks.com

Mythology Photography **Fiction** Fishing Christianity **Art** Cooking Essays Buddhism Freemasonry Medicine **Biology** Music **Ancient Egypt** Evolution Carpentry Physics Dance Geology **Mathematics** Fitness Shakespeare **Folklore** Yoga Marketing **Confidence** Immortality Biographies Poetry **Psychology** Witchcraft Electronics Chemistry History **Law** Accounting **Philosophy** Anthropology Alchemy Drama Quantum Mechanics Atheism Sexual Health **Ancient History** **Entrepreneurship** Languages Sport Paleontology Needlework Islam **Metaphysics** Investment Archaeology Parenting Statistics Criminology
Motivational

First American Edition.

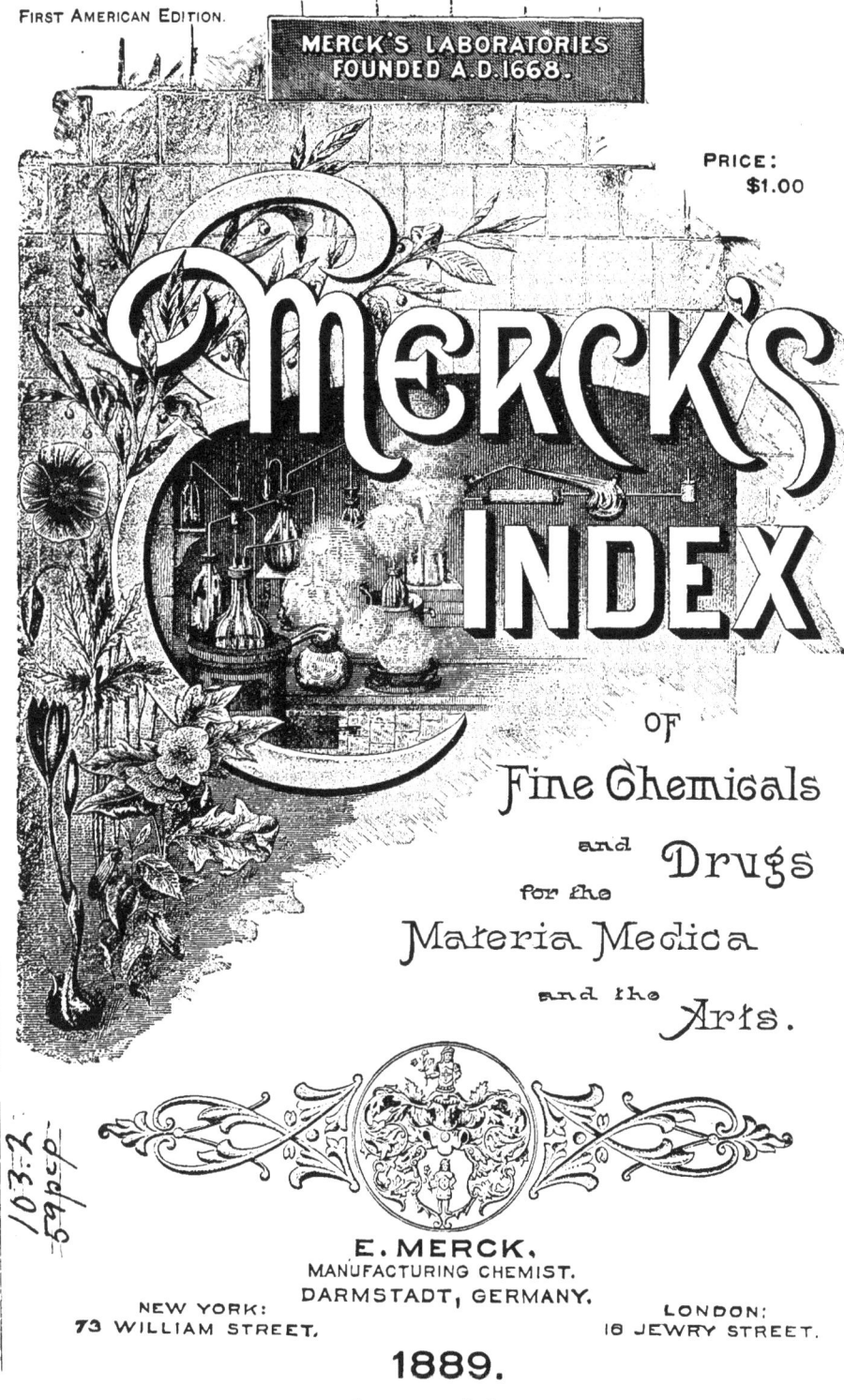

MERCK'S LABORATORIES FOUNDED A.D. 1668.

PRICE: $1.00

Merck's Index

of Fine Chemicals and Drugs for the Materia Medica and the Arts.

E. MERCK,
MANUFACTURING CHEMIST.
DARMSTADT, GERMANY.

NEW YORK:
73 WILLIAM STREET.

LONDON:
16 JEWRY STREET.

1889.

Copyright by E. Merck.

MERCK'S
— OWN —
AMERICAN HOUSE

IS LOCATED IN

NEW YORK CITY,

No. 73 WILLIAM STREET.
[P. O. Box 2649.]

THEODORE WEICKER,
Empowered Attorney and General Business Manager for E. Merck in the U. S.

※

E. MERCK,

| NEW YORK, | DARMSTADT, | LONDON, |
| *U. S. A.* | *Germany.* | *England.* |

Manufacturing Chemist and Pharmaceutist,
AND
Purveyor to the Materia Medica of all Countries.

MERCK'S LABORATORIES AT DARMSTADT WERE FOUNDED IN THE YEAR
===1668.===

MERCK'S INDEX

— OF —

Fine Chemicals and Drugs

FOR THE

MATERIA MEDICA

AND THE

ARTS.

COMPRISING A SUMMARY OF
WHATEVER CHEMICAL PRODUCTS ARE TO-DAY ADJUDGED AS BEING USEFUL
IN EITHER MEDICINE OR TECHNOLOGY.

WITH AVERAGE VALUES AND SYNONYMS AFFIXED.

A GUIDE

For the Physician, Apothecary, Chemist, and Dealer.

BY

E. MERCK.

1889.

Entered, according to Act of Congress, in the year 1889, by
E. MERCK,
in the Office of the Librarian of Congress, at Washington, D. C.

NEW YORK, *January*, 1889.

To the Members of the *Medical* and *Pharmaceutical* Professions of America.

Dear Sirs:—

In looking back upon the line of generations during which my Home Office and Laboratories at Darmstadt, Germany, have been in existence, I find that yonder Office has, for many years past, held agreeable relations with you, gentlemen of both professions in America, through the intermediation of your Importers and Drug Merchants. I find, furthermore, that those relations have become widened in extent and deepened in reciprocal regard, with unfailing constancy, as year after year wore on.

This was made manifest to me, from time to time, in many different ways; among others,—by numerous requests from distinguished members of your professions, to the effect that I would provide a more convenient avenue of mutual communication between us.

The continued recurrence of these requests, and the multiplying number of the sources whence they came, finally caused me to accede to them, by establishing a House of my Own in America,—which was opened in *February of 1887*.

That action of mine, however, was *in no wise* inspired by any distrust or unfriendly sentiment, on my part, toward the able and respected merchants who always have been, and still are, the intermediaries of your intercourse with me. They have not in the least changed their position in this regard; with the sole exception that, instead of being obliged, heretofore, to send their orders for my products to my Darmstadt office, they now obtain their supplies directly and promptly from my American warehouse, which is more readily accessible to them. Hereby the course of trade in these chemicals is not altered in any other wise than that of added ease, promptness, and certainty of execution. Thus, my business relations with the American Wholesale Drug and Chemical Trade remain precisely as they were before the establishment of my own General Depot at New York. My moral relations with you, gentlemen of both professions, will, I am bold to hope, likewise remain as heretofore,—those of mutual esteem and confidence; with the modification, perhaps,—resulting from the comparative nearness of my American establishment to you and your purveyors,—of making many of you, as well as of them, still better acquainted with the *vastly comprehensive extent* of the full line of my products, numbering to-day upwards of 5,000 medicinal, analytical, and technical Chemicals; thus embracing about every purely chemical compound or derivative, and most of the pharmaceutical preparations, at present employed in Medical Art.

The present volume contains an alphabetically arranged List of those of my products which are, at the present day, dealt-in by the principal Drug and Chemical Warehouses in all parts of the world; *added to which* are about a dozen preparations mostly made under patent restrictions by other makers exclusively, and which, on account of their excellence and importance, have been received into this "Index."

The most vital interests of your patients, gentlemen physicians!—and of your customers, gentlemen of the pharmaceutical profession!—depend, as you are well aware, on the reality of the Presumed Purity, of the Prescribed Strength, and of the Correct Condition of the materials employed in filling prescriptions. Your well-founded confidence in the Standard and Reliable Brand of "Merck" may, in many cases, *where you have not found an equally certain* preparation from other sources, cause you *to specify* "Merck's" in your prescriptions to be filled by your Dispensing Pharmacist, or in your orders sent to your Wholesale Dealer.

Such specifications can now be obeyed within a comparatively brief time, when not instantly, *by* every *Apothecary,—or by* every *Drug and Chemical Merchant, respectively,*—throughout *the length and breadth of our States and Territories;* for, whenever a substance specified as "MERCK'S" should not be thus in stock at the moment when first required, the next *return mail from New York* will, as a rule, bring it whithersoever desired! This is the great achievement gained for the friends of my Brand on this Continent by the establishment of my American Branch: that almost anything likely to be desired from the vast arsenal of the Materia Medica can now be obtained at very short notice from my well-stocked New-York warehouse; whereas, formerly, many weeks may have elapsed before a given special order could be filled *via* Atlantic steamer.—For it must be borne in mind that, in my house in this city, I keep a full line of my own products, consisting not only of those rarer and difficultly obtainable Botanical Derivatives, mostly known as *Alkaloids, Glucosides,* or *Resinoids,* (which constitute, it is true, a *special* and *eminent* province of my Laboratories),—but likewise, as above indicated, of all the *Metallic Salts* and *Synthetical Organic Compounds,* etc., employed in Modern Medicine;—besides the most important of the *regular Pharmaceutic Preparations* (Balsams, Essences, Extracts, Juices, Oils, Resins, Solutions, Spirits, Syrups, Tinctures, Waters, etc.);—added to which are all the Laboratory Reagents employed by Analytical Chemists, and a great number of the *Finer Grades of Technical Chemicals* (Acids and other Solvents, Anti-Ferments, Detergents, Mordants, Pure Metals, etc.).

Furthermore, I would beg leave to direct the attention of Physicians and Druggists to the fact that all these preparations, whenever "Merck's" Brand is called for, can be furnished by every Drug and Chemical Warehouse of the United States and Canada, *in the Original Package* and under the Original Label and Seal of my Darmstadt Laboratories,—*be the package of any size,* small or large, that may be desired.

I would earnestly entreat my friends, throughout both professions, to insist rigidly that Merck's Chemicals be furnished to them, by dealers, in the *original* packages as above described. If any dealer refuses, or professes to be unable, to thus furnish them—after being allowed a reasonable lapse of time for correspondence with my New-York Office—I will be thankful to parties thus disappointed if they will communicate full particulars to me, at New York City (73 William Street, or P. O. Box 2649), and *I will in each case endeavor to procure the prompt satisfaction of the demand made.*

I shall also feel pleased, at all times, to give to professional gentlemen *any other desired Information at my command.*

Quite a number of inquiries, however, such as come to me by each mail in great numbers, might have been averted if the inquirers had read a Monthly Publication issued by me, entitled: "Merck's Bulletin—a periodical record of New Discoveries, Introductions, or Applications of Medicinal Chemicals." That journal is issued *exclusively for the purpose* of informing professioual men on what may be *of actual interest* to them in the field of chemical, physiological or therapeutical discovery as to Chemico-medicinal Preparations.—"MERCK'S BULLETIN" is edited *in the briefest possible form*, leaving aside all speculative ventures of opinion, and confining itself to established facts. It is further edited *without deference to Merck's or any one else's business interests,*—simply describing Things that are New and Interesting, without any regard whatever to their origin, sale, or trade-connection.

One remark may be needed by my professional friends, as to the Price-notes placed opposite the names of most substances in the following List. *Those Price-notes are* not *intended to give this work the character of a commercial or business Price-list.* The prices of most of the articles enumerated are, in the nature of the market, variable; and the *sole purpose* of inserting such price-notes here is, therefore, to give Physiciaus and Apothecaries a somewhat approximative idea as to *Average Market Values;* so as to serve as an occasionally convenient guide in calculating the cost of various medicines, and, consequently, in some cases, to assist in determining their choice, when there may be several substances of like mode of action to choose from, and when the item of cost may have to be a factor in the selection.

It will be understood that the Values stated are based on the average rates which the Retail Druggist is expected to pay his purveyor; and that, couse- quently, they will form a basis *only to the Apothecary or to the Dispensing Physician* for the calculation of his own expenditure.

The *Ruling in the blank columns* after the price-notes is intended for the insertion of private notes regarding the stated articles.—The cross- ruling at the end of each alphabetical division may serve to allow new articles to be added.

The *English Nomenclature* and *Orthography* hereinafter followed, for the designations of chemical compounds, are, in the main, those adopted by the Chemical Society of England, and by most of the modern text-books and treatises on chemistry, both in England and the United States.—For instance, the termination *"ine"* is reserved strictly for only two classes of bodies: *Elements* (Chlorine), and *Alkaloids* or *other* non-metallic *Bases* (Strychnine; Hydroxyl-amine); while all *Glucosides, Resinoids, Amarulents, Proteids,* or other *Neutral* or prevalently *Acid* bodies drop that *"e"* (Strophanthin; Agaricin; Euonymin; Chondrin; Tannin).—*Hydrocarbons* of the *Aromatic* Series end in *"ene,"* supplanting "ol" or "in" or "en" (Benzene [not "Benzol"]; Naphthalene [not "Naphthalin"]; Stilbene [not "Stilben"]);—those of the *Fatty* Series in *"ane"*—not "an"—(Methane). [Some *Esters* likewise end in *"ane"* (Ur-ethane), and some in *"in"* (without final *e*)—(Stearin)].—The termination *"ile"* carries the mute *e* (Nitrile); the termination *"yl"* does not (Acetyl).—*Alcohols* (so-called Hydroxyl-derivatives of Hydrocarbons) *do not* add a mute *e* to the termination *"ol"* (Carbinol), while the *other* compounds ending similarly take the *e* for distinction (Indole). [With some Alcohols, the termination *"in"* has become so firmly established in current usage, that this was recognized in the List; as, f. i.,—" Glycerin = Glycerol." Some of the *higher* (poly-hydric or poly-valent) *Alcohols* of the Fatty Series have been given under the *distinctive* termination of *"it,"* with other recognized forms added ("Mannit' = Mannitol = Mannol"); while the termination *"ite"* has been reserved wholly for *Salts* of the weaker Acid-forms (—Nitrite) and *Native Minerals* (Pyrolusite).]—"Aldehyd" has been deprived of the final *e* appended to it by many authors, as being more exactly in accordance with its etymology of

"*Al*[cohol] *dehyd*[rogenatus]."—These are some of the principal *Orthographical* points on which various authors are still in the habit of differing. —As to *Nomenclature proper*, there will, I presume, be no difficulty of understanding, inasmuch as the system hereinafter used is one that has been taught in our schools, in substantially the same form, for nearly a generation past.

In connection herewith I would say that quite a great deal of labor has been bestowed, in arranging the matter of the book, on the introduction of a pretty full array of Synonyms (embracing both *popular* or *trade,* and *alchemistic* or so-called *magistral* designations).—I was originally loth to call the products here listed by any other than their properly (and when so: officially) received chemical appellations,—intending to add only a few of the pharmacopeial designations in cases where these differed from the former. But such floods of both orders and inquiries poured in upon me equally from Trade and from Professional quarters, *using the most various designations for same objects*, that I found myself perforce compelled—if I meant to accommodate the mass of my readers—to receive into the List a number of names deemed quite obsolete by me at the first planning of this work.

But, whichever the "odd names" thus received may be,—the substance in question is *invariably listed under a proper chemical name* also, and is, as a rule, *detailed* and *priced* there! (In no case is a substance detailed or priced in two or more places in the List, but *always*— if at all—*only* in the place pointed-to by the words "*see* ——," or "*see under* ——."—Thus: the trade-names "Vitriol, blue," and "Copper Vitriol" are both found in the List in their respective alphabetic places; but, after both, the reference-remark points to "Copper, sulphate, neutral"; *where alone* the Descriptions and Market-values of its different forms and qualities are stated.) In a very few instances, the money-value of a substance is stated *after a name quite different from any of its proper chemical designations;* such departure is then always due to a differing pharmacopeial (U.-S.) nomenclature. (For example: "Calcium, oxide," is referred to "Lime," because the U.-S. Pharmacopœia calls it "Calx = Lime.")—Whenever a substance is here listed under a name *deviating* from the English form of its U.-S. pharmacopeial Latin name, the latter is *always added* in parentheses, and is also repeated (in English) in its proper alphabetic place, as a *Synonym*. (For example: "Mercury, bichloride," has after it the parenthesis "Hydrargyri chloridum corrosivum," *and* is also listed under the synonym: "Mercury, chloride, corrosive.")—In a few other instances, when substances had to be referred, for their quality-standard or mode of preparation, to some *Foreign* Pharmacopœia, their Latin synonyms, when given in such connection, are formed according to the system of nomenclature of *that* particular work. (For example: "Antimony, oxide, precipitated," will be found described in parentheses, first, by its exact chemical designations: "Antimonious oxide—Tri-oxide";—then by its U.-S. pharmacopeial name: "Antimonii oxidum";—and then again by one of its foreign pharmacopeial names: "Stibium oxydatum præcipitatum.")

When a complicated compound may as likely be sought-for under its *rational chemical name* as under its *empirical chemical name*, both are listed. (Thus: "Urea" = "Carb-amide"; "Pyro-catechin" = "Di-oxybenzene, ortho-.")

I sincerely trust the book may be a Welcome Visitor not only to whomever it calls upon; but may prove so *useful* as to be asked to "come again."

E. MERCK.

> The ORIGINAL DOCUMENT, of which the subjoined text contains a literally identical reproduction, is to-day preserved in the GRAND-DUCAL STATE ARCHIVES at DARMSTADT, Germany.—The meaning of the ancient text, dated July 10th, 1682, is that of a GOVERNMENT CHARTER, or LETTERS-PATENT, confirming and continuing, to GEORGE FREDERICK MERCK, the CHARTER or GRANT OF LICENSE conferred upon JACOB FREDERICK MERCK IN THE YEAR 1668, BY THE LANDGRAVE OF HESSE: LUDWIG THE SIXTH, —for the maintenance of a PHARMACEUTIC ESTABLISHMENT by said Merck.— The Establishment referred-to has now been in the possession and under the direction of the MERCK FAMILY FOR 221 YEARS, and has by them, in the meantime, been developed into the immense complex system of MANUFACTURING LABORATORIES, to-day known as
> "MERCK'S DARMSTADT CHEMICAL WORKS."

Copia copiae.

Von GOTTES Gnaden Wir Elisabetha Dorothea,

Landgräfin zu Hessen, Fürstin zu Herßfeld, geborene Herzogin zue Sachsen, Jülich, Cleve und Berg p. Gräfin zue Catzenelnbogen, Dietz, Ziegenhain, Nidda, Schauenburg, Ysenburg und Büdingen p. Wittib, Vormünderin und Regentin, Thun kund und bekennen in Vormundschaft Unseres freundl. geliebten ältisten annoch Minder Jährigen Sohns, Landgraf **Ernst Ludwigs** zu Hessen p. hiermit, Alß Sr. Ldl. hochseel. Herr Groß Vatter, Weyland Herr Landgraf **Geörg** zu Hessen p. Weyland Johann Samuel Böcklern im Jahr 1654 und folgends nach dessen Absterben, Unsers nunmehr in Gott ruhenden Herrn und Ehemahls, Weyland Herrn Landgraf **Ludwigs**, des Nahmens der Sechsten zu Hessen p. Lbdl. im Jahr 1668. Jacob Friederich Mercken von Schweinfurt, die Gnad gethan, und ihnen eine Apotheck allhier aufzurichten und *respective* zu *continuiren*, ein *Privilegium* und Verwilligung ertheilet; Und dann seithero Beedes erwehnter Johann Samuel Böckler und Jacob Friederich Merck verstorben, und Uns darauf jetztgedachtes Jacob Friederich Merckens Vetter, **Georg Friederich Merck**, umb ertheilung solches Apothecker *Privilegii* auf ihne unterthänigst gebetten; Und Wir ohne das, zu desto mehrer erhaltung der *Medicorum* und *Patient*en *Libertät* und Vermeydung sonstschädlichen *Monopol*-Wesens, ohne das gern sehen, daß zwey wohlbestelte Apothecken allhier seyen und erhalten werden; Daß Wir, so gestalten sachen und Umbständen nach, in sothanes sein Geörg Friederich Merckens Suchen gnädigst gewilliget, Thun dasselbe auch hiermit und in Kraft dieses, in der Besten und Beständigsten form, als es von Rechts- und Gewohnheit wegen geschehen soll, kann und mag, Und soll er Geörg Friederich Merck sich hingegen der Fürstlichen Hessischen Apothecker Ordnung jederzeit gemees verhalten, ehist die gewöhnliche pflichten Leisten, zumahl aber seine Apotheck nicht weniger, als der andere Apothecker *Scipio*, die seinige, soweit es nicht schon geschehen ist, dergestalt mit guten frischen, zu ein- und andern *Curen* dienlichen heylsamen *Medicamentis* und Wahren, also genugsamlich versehen, und damit fort und fort würklich *continuiren*, daß kein Mangel erscheine und also allhier zwey rechtschaffene wohlbestelte, zum wenigsten *in qualitate*, weil es etwann *in quantitate* nicht allezeit wohl geschehen könnte, einander gleichstreichende *Corpora* seyen, wie auch die *Medicamenta* dem Armen sowohl als dem Reichen, beedes in der Gütigkeit und Billichen Leidlichen, und zum wenigsten in dem zu Franckfurt vor Meß- zu Messen üblichen *tax* und Preiß /: es were dann daß Wir in etlichen Stücken ein sonderbare *tax* Ordnung ausgehen ließen :/ geben und folgen lassen, Jumasen Wir die *Visitation* Besagter Apothecken durch Unsere darzu *Deputirte* Rhäte auch *Medicos*, und wen Wir sonsten noch weiter darzu *deputiren*, nach und nach zu verfügen, nicht unterlassen werden; Befehlen und verordnen darauf und wollen, daß wieder dieses *Privilegium* und Vergünstigung nichts nachgesehen, noch verhenget, sondern derselbe vielmehr, so lang er sich vorgeschriebenermaßen und sonsten der Gebühr verhalten wird, darbey gehandhabt und darwider nicht beschweret werden soll, treulich und ohne Gefährde; Uhrkundlich Unserer Aigenhändigen Unterschrift und hierauf gedruckten Fürstlichen *Secrets*, *Datum* —

Darmstadt am 10ten *July* anno 1682.

Elisabetha Dorothea Landgräfin zu Hessen.
(**L. S.**)

"SUUM CUIQUE."

The list herewith submitted, of a few of the HONORABLE AWARDS extended to the firm of **E. MERCK**, embraces, by the desire of the House, but A NUMERICALLY SMALL FRACTION of such awards received during the time from 1830 to 1883; the balance not enumerated may be covered by the remark that **E. MERCK** NEVER EXHIBITED HIS PRODUCTS ON ANY PUBLIC OCCASION WHATEVER, WITHOUT THEIR ELICITING A TOKEN OF ESPECIAL DISTINCTION AND HONOR.

THEODORE WEICKER,
Manager in the U. S. for E. MERCK.

Among the **AWARDS** received by **E. MERCK**, of Darmstadt, are the following:

1830: Gold Medal: "For the Relief of Mankind." — Pharmaceutical Society of PARIS, (France). Competitive Exposition.

1853: Medal and Special Approbation: "For Specimens of Alkaloids." — Exhibition of the Industry of All Nations,— NEW YORK.

1861: Gold Medal and Diploma. — Industrial Exposition for the Grand Duchy of Hesse,—DARMSTADT.

1862: Medal: "Honoris Causa." — World's Fair,—LONDON, (England).

1864: Award: "Beyond Competition" (PRIX HORS LIGNE): "Numerous and varied collection of Alkaloids and very rare products; Physiological Preparations of high interest and very difficult to obtain in any appreciable quantity." — Pharmaceutical Congress of France. Hygienic and Pharmaceutic Exposition, STRASSBOURG.

1867: Gold Medal: "Chemical Preparations; Quinine Salts; Alkaloids." — Universal Exposition,— PARIS, (France).

1873: Medal of Progress and Diploma. (The Highest Award.) — World's Exposition,— VIENNA, (Austria).

1876: The Great Prize Medal and Diploma. — Industrial Exposition for the Grand Duchy of Hesse,—DARMSTADT.

1879: "First Award." — International Exhibition,— SYDNEY, (Australia).

1880: Gold Medal and Diploma: "A Fine and Vast Collection of the Rarest Alkaloids and their Salts." — Medical Assoc'n of Italy. Ninth Convention, Third Exposition, GENOA.

1880: Gold Medal: "Vitam Excolere per Artes." — International Exhibition,— MELBOURNE, (Australia).

1883: The Diploma of Honor. — International Exposition,— AMSTERDAM, (Holland).

MERCK'S CHEMICALS are to be obtained through the *Wholesale and Jobbing Drug Trade* in all parts of the United States, in **UNBROKEN ORIGINAL PACKAGES** (of any desired size!) under the *Genuine Darmstadt Seal and Label.*

☞ *Whenever difficulty is experienced in thus procuring them, relief will be had by sending prompt notification to:*   **E. MERCK**, NEW YORK CITY. (P. O. Box 2649.)

☞ Table of Abbreviations, see page 156.	Containers incl.			
Absinthin (Absynthiin)................	15 gr. .75			
Acetal (Di-ethyl-acetal), commercial......	oz. .75			
" pure............................	oz. 1.00			
Acetal, di-Methyl-, see Di-methyl-acetal..				
Acet-amide	oz. .65			
Acet-anilide,—medicinal,—see **Antifebrin**.....				
" mono-bromated, see Brom-phenyl-acet-amide, mono-...................				
Aceto-acetic Ester, see Ethyl, aceto-acetate				
Acetone (Di-methyl-ketone), [so-called Pyro-acetic "Ether" or "Spirit"]...	lb. 1.10			
" chem. pure,—boiling-point 56-58° C [132.8-136.4 F]..................	lb. 1.50			
Aceto-nitrile, see Methyl, cyanide........				
Aceto-phenone, see **Hypnone**................				
Acet-phenetidin, para-, see Phen-acetin..				
Acetum concentratum, purum; and, purissimum, Ph. G. II;—see Acid, acetic, pure,-solution; and, ch. pure,-solut.				
" **plumbicum** (*Saturni*), see Solutions: Lead acetate, basic, *U. S. Ph.*........				
" **pyrolignosum** rectificatum, Ph. G. II, see Acid, pyro-ligneous, purified....				
Acetyl Chloride.........................	oz. .50			
Acetylene-urea (Acetylene-carbamide)....	15 gr. 1.00			
Acid, acetic, pure,-solution, (Acetum concentratum purum),—sp. gr. 1.04..	lb. .50			
" " chem. pure,-solut., (Acetum concentr. puriss., Ph. G. II),—sp. gr. 1.04, [30% of $C_2 H_4 O_2$]	lb. .55			
" " chem. pure,—*U. S. Ph.*,—sp. gr. 1.048, [36%]	lb. .60			
" " pure,—sp. gr. 1.060 } [50% of	lb. .50			
" " ch. p.,—sp. gr. 1.060 } $C_2 H_4 O_2$]				
N. B.—The "*chem. pure,*— sp. gr. 1.060,"—is indifferent to Permanganate of Potassium.	lb. .60			
" " glacial,—*U. S. Ph.*,—[99%];—dissolves Oil of Lemon in any proportion...............	lb. .85			
" " " —exactly acc. to Ph. G. II, [96% of $C_2 H_4 O_2$]	lb. .85			
" " " —[85%]; dissolves Oil of Cloves	lb. .60			
" " anhydrous	oz. .50			
" " pyro-ligneous, rectified, see Acid, pyro-ligneous, purified........				
" aconitic,—identical with *Achilleic* acid..	15 gr. .25			
" æthyl-malonic, see Acid, ethyl-malonic.				
" agaric (agaricic, agaricinic), see Acid, laricic..........................				
" aloe-resinic,—according to Mulder.....	15 gr. .25			
" aloetic (aloetinic)....................	15 gr. .25			
" amido-acetic (amido-glycollic), see Glycocoll...........................				
" amido-caproic, see Leucine...........				
" amido-ethyl-sulphonic, see Taurine...				
" amido-succinic, see Acid, asparagic....				
" amygdalic, (*not* Amygdalinic acid!), see Acid, mandelic....................				
" amylic, see Acid, valerianic...........				
" anacardic............................	15 gr. .50			
" anemonic............................	15 gr. 1.75			
" anilotic (anilotinic)...................	15 gr. .25			
" anisic, cryst.	15 gr. .25			

☞ **When ordering, specify: "MERCK'S"!**

MERCK'S INDEX.

	Containers incl.			
Acid, antimonic, anhydrous, see Antimony, oxide, white, *true*, (Pent-oxide)......				
" antimonious, anhydrous, see Antimony, oxide, precipitated, pure, (Tri-oxide).				
" arabic (arabinic) [gummic], see Arabin				
" arsenic (arsenicic), hydrated,—soluble,—[Tetra-hydrated Arsenic Pent-oxide; Hydrated Tri-hydric Arseniate — $H_3 AsO_4 \cdot \frac{1}{2} H_2 O$],—pure......	lb. 1.00			
" " dry (anhydrous), — [Arsenicic Anhydride, Arsenicic Oxide; Arsenic Pent-oxide — $As_2 O_5$], — commercial...............	lb. .90			
" arsenious (arsenicous), anhydrous, — [Arsenious Anhydride, Arsenious Oxide; Arsenic Tri-oxide; so-called "White Arsenic," — Resublimed "Flowers of Arsenic"], — pure, *lumps*; — (*Vitreous* Arsenic, Arsenic-glass)....... *conforming to U. S. Ph. and Ph. G. II.*	lb. 1.00			
" " do.,—pure, powder....	lb. 1.50			
" asparagic (asparaginic, aspartic) [amidosuccinic]..........................	15 gr. .35			
" atropic............................	15 gr. 1.00			
" benzoic, from Siamese Benzoin-resin; sublimed,–Ph. G. II *Flowers of Benzoin.*	lb. 8.50			
" " fr. Benzoin-resin; sublimed, — U. S. Ph. and Ph. G. II..	lb. 7.50			
" " fr. Benzoin-resin; sublimed, perf. white...............	oz. .20			
" " from Benzoin-resin; wet process, cryst..........................	oz. .30			
" " from Toluol....................	lb. .85			
" " from urine; sublimed..........	lb. 2.25			
" " " " resublimed, perfectly white, chem. pure..	lb. 3.00			
" bi-chlor-acetic, see Acid, di-chlor-acetic.				
" boric (boracic), crude, cryst..........	lb. .40			
" " ch. pure, perf. white, cryst., —U. S. Ph............ *conform. to Ph. G. II.*	lb. .60			
" " ch. pure, perf. white, powder	lb. .65			
" " " " " "impalp. pwd.	lb. .75			
" " pure, perf. white, cryst........	lb. .50			
" " " " " powder.........	lb. .55			
" " " " " impalp. powder.	lb. .60			
" " " fused....................	lb. 2.00			
" " glycerolate (glycerite) of, see Boro-Glycerin, *dry*........				
" boro-benzoic	oz. .50			
" " -citric.........................	oz. .25			
" " -hydrofluoric..................	oz. .35			
" " -salicylic......................	oz. .75			
" " -wolframic (boro-tungstic).......	oz. 1.75			
" bromic,—sp. gr. 1.12................	oz. 1.00			
" bromo-acetic........................	oz. 1.75			
" bursic.—The active principle of Bursa pastoris, (Capsella B. p.), [Shepherd's purse].-(Highly efficient hemostatic.)				
" butyric, normal, concentrated, — [abt. 60-65%].............	lb. 1.75			
" " " chem. pure............	lb. 4.00			
" " Iso-............................	oz. 1.00			
" cacodylic (kakodylic) [di-methyl-arsinic].—Also called. "Alkargen" (*not* to be confounded with "Alkarsin"!).				
" cahincic (caincic), [Cahincin]..........				

☞ **When ordering, specify: "MERCK'S"!**

	Containers incl.		
Acid, camphoric, —melt.-point 178° C [352.4 F].—(Recently introduced into therapeutics as an inhalant in diseases of the air-passages; also, as a surgical aseptic, etc.)	oz. 1.00		
" capric (caprinic) [rutic]	oz. 4.50		
" capronic (caproic), pure	oz. 1.25		
" caprylic	oz. 4.00		
" carb-azotic, see Acid, picric			
" carbolic (phenic, phenylic), chem. pure, loose crystals,—[Absolute Phenol; so-called "Hydrate of Phenyl"],—melt.-point 40° C [104 F],—U. S. Ph.—As to purity, both this grade and the following correspond to: } Ph. G. II.	lb. 1.00		
" " pure, cryst., fused, white, — melt.-point 35° C [95 F] }	lb. .60		
" " liquid, brown, [ab. 90%],—Ph. G. II			
" " " crude I, [50-60%] } U. S. Ph.			
" " " " II, [30%]			
" " " III			
" " solution [90%] in Glycerin,—(Phenol-Glycerin), [Glycerolate (Glycerite) of Carbolic acid];—for medical use	lb. 1.25		
" " iodized, (Iodized Phenol)	oz. 2.00		
" carminic, chem. pure	oz. 2.00		
" carthamic, so-called, see Carthamin			
" caryophyllic, formerly so-called, (Eugenic acid), see Eugenol			
" catechuic, see Catechin			
" catechu-tannic, chem. pure	oz. 2.00		
" **cathartic** (cathartinic), [not identical with Cathartin, — which see also!]	oz. .75		
" " pure	oz. 1.00		
" cerebric (cerebrinic)	15 gr. 2.00		
" cerotic (cerotinic)	15 gr. .75		
" cetraric, see Cetrarin			
" cheno-cholic (cheno-cholinic)	15 gr. 1.00		
" chinic, see Acid, quinic			
" chino-picric, see Acid, quino-picric			
" chinovic, see Acid, quinovic			
" chloric,—sp. gr. 1.12	oz. .25		
" " per-, see Acid, per-chloric			
" chloro-acetic.—(An escharotic.)	oz. .60		
" chloro-chromic, anhydrous, (Chlorochromic Anhydride), see Chromium, di-oxy-di-chloride			
" chloro-nitrous. (chlor-azotic), see Acid, nitro-hydrochloric, U. S. Ph.			
" choleic (choleinic), see Acid, tauro-cholic			
" cholic (cholalic), cryst.	15 gr. .75		
" " amorphous	15 gr. .60		
" choloidic (choloidinic)	15 gr. .50		
" **chromic, cryst., chem. pure,—absolutely free from Sulphuric acid.**—(Solely a Chromic acid possessing this qualification is fit for use as an escharotic.)	oz. .30		
" do.,—same as above—in pencil form	oz. 1.00		
" chromic, pure, cryst.,—U. S. Ph.	oz. .18		
" " commercial	lb. .75		
" chromo-nitric	oz. .25		
" chrysammic (chrysamminic)	15 gr. .50		
" **chrysophanic.**—(so-called),—medicinal,—see **Chrys-arobin**			
" " —true,—(Rheic acid), see Rhubarb constituents: Rhein			

☛ When ordering, specify: "MERCK'S"!

		Containers incl.		
Acid, cinnamic (cinnamylic), chem. pure		oz. 1.50		
" " crude		oz. 1.00		
" citric, colorless, cryst.		lb. 1.25		
" " " powder		lb. 1.35		
" " " pure, cryst. { free fr. Lead		lb. 1.35		
" " " " powd. }		lb. 1.45		
" " " ch. pure, —U. S. Ph.,—cryst., { absolutely pure and conforming to Ph. G. II.		lb. 1.50		
" " " ch. pure, powder		lb. 1.60		
" copaivic, amorphous		oz. .75		
" " cryst., (Meta-copaivic acid)		oz. .40		
" " crude, see Resins: Copaiva				
" cresotic (cresotinic)		oz. .50		
" cresylic, (Cresol)		oz. .40		
" crotonolic,(not Crotonic, but Tiglic [Methyl-crotonic] acid!)		15 gr. .60		
" cubebic		15 gr. .60		
" cumarylous (coumarylous), [Cumaric Anhydride], see Cumarin				
" cuminic, cryst.		15 gr. .40		
" cyan-uric (tri-cyanic), cryst.		15 gr. .35		
" di-chlor-acetic (bi-chlor-acetic), pure		oz. 1.50		
" di-iod-salicylic		15 gr. .50		
" di-methyl-arsinic, see Acid, cacodylic				
" di-methyl-nor-opianic, see Acid, opianic				
" di-methyl-proto-catechuic, see Acid, veratric				
" elaic (elainic—not elaidic, elaidinic!), see Acid, oleic				
" elaidic (elaidinic).—An isomeric modification of Oleic acid		15 gr. .75		
" **elateric, anhydrous, see Elaterin Merck, cryst.**				
" ergotic (ergotinic),—according to Zweifel N.B.—See, also: Acid, sclerotic, etc.		15 gr. 2.50		
" ethyl-di-acetic, see Ethyl, aceto-acetate				
" ethyl-malonic		15 gr. .50		
" ethyl-sulphurous (not: ethyl-sulphuric!), see Acid, sulpho-vinous				
" eugenic, (formerly called "Caryophyllic acid"), see Eugenol				
" filicic, (Filicin)		15 gr. .50		
" formic (formylic), pure,—Ph. G. II,— sp. gr. 1.060 ... [25% CH_2O_2]		lb. 1.50		
" " " pure,—sp.gr. 1.120, [50% "]		oz. .25		
" " " " 1.150, [65% "]		oz. .30		
" " " " 1.180, [80% "]		oz. .35		
" " " " 1.200, [90% "]		oz. .40		
" " " " 1.220, crystallizable, [100% CH_2O_2]		oz. .65		
" frangulic (frangulinic)		15 gr. .50		
" fumaric		15 gr. .30		
" gallic, cryst.,—U. S. Ph.		lb. 1.25		
" gaultheric (methyl-salicylic), so-called, see Methyl, salicylate				
" gentianic (gentisic), see Gentisin				
" glyco-cholic		15 gr. .75		
" gummic (arabic), see Arabin				
" gynocardic		oz. 1.50		
" **hippuric, cryst.**		oz. 1.50		
" hydrobromic,-sp.gr.1.49,[abt.48% HBr]		lb. 2.50		
" " sp. gr. 1.38 [" 40% "]		lb. 1.75		
" " " 1.27 [" 30% "]		lb. 1.50		
" " acc. to Fothergill. [" 12% "]		lb. 1.00		
" " diluted,—U. S. Ph.,—sp.gr. 1.077, [10%]		lb. .75		
" hydrochloric (muriatic), pure,—sp. gr. 1.190, [38.5% H Cl]		lb. .50		

☞ When ordering, specify: "MERCK'S"!

		Containers incl.	
Acid, hydrochloric, —(as above!);—sp. gr. 1.16, [31.8% H Cl]; conforming to U. S. Ph. and Ph. Brit......		lb. .40	
" " —sp. gr. 1.124, [25% H Cl]; conforming to Ph. G. II..........		lb. .38	
" hydro-cinnamic (hydro-cinnamylic)....		15 gr. .50	
" hydrocyanic (prussic), diluted, — U. S. Ph.,—abt. 2% of CNH............		oz. .17	
" hydrofluoric, fuming.............		oz. .50	
" hydro-iodic (hydriodic),—sp. gr. 1.50, [47% HI].................		oz. .60	
" " sp. gr. 1.70, [57% HI].........		oz. .70	
" hydro-silico-fluoric,—sp.gr. 1.060,[9°Bé]		lb. .60	
" " sp. gr. 1.157, [20° Baumé].......		lb. 1.00	
" hyo-cholic (hyo-cholalic).............		15 gr. .75	
" hyo-glyco-cholic		15 gr. .50	
" hypo-phosphórous,—sp. gr. 1.15		oz. .25	
" **ichthyol-sulphonic,** see under Ichthyol preparations...........................			
" inosinic			
" iodic, cryst.		oz. .80	
" " anhydrous		oz. 1.00	
" iodo-salicylic		oz. 3.00	
" " -tannic, solution...............		lb. .75	
" iso-butyric, see Acid, butyric, Iso-.....			
" iso-valeric,—various kinds,—see Acid, valerianic			
" kakodylic, see Acid, cacodylic........			
" kinic; kino-picric; kinovic;—see Acid, quinic; quino-picric; quinovic......			
" kresotinic, } see Acid, { cresotic.......			
" kresylic .. } { cresylic			
" **lactic, white,** (Iso-lactic [Fermentation-lactic] acid),—optically inactive,—**sp. gr. 1.21,**—U. S. Ph...	conform. in purity to Ph. II	lb. 1.80	
" **do., do.,**—do. do., - **sp. gr. 1.16** ...		lb. 1.50	
" lacto-arsenious, see Arsenic, lactate....			
" laricic (agaric, agaricic, agaricinic), — from White Agaric—Fungus laricis; - [not identical with Larixinic acid, from Pinus larix!];—(furthermore: not identical with Agaricin,—which see also!)		oz. 4.00	
" lithic, see Acid, uric...............			
" malic (oxy-succinic),—optically active, —pure		oz. .90	
" malonic		oz. 2.00	
" mandelic (phenyl-glycollic), [Amygdalic —not Amygdalinic!— acid].........		15 gr. .50	
" margaric (margarinic)...............		oz. 3.50	
" meconic, cryst.		oz. 3.00	
" mellitic (mellic)...................		15 gr. .75	
" methyl-crotonic (tiglic), see Acid, crotonolic			
" methyl-proto-catechuic, see Acid, vanillic			
" methyl-salicylic (gaultheric), so-called, see Methyl, salicylate			
" methyl-tri-hydro-oxy-quinoline-carbonic,—[$C_{11}H_{13}O_3N$—acc. to Nencki, of Basle],—Sodium-salt of,—see Thermifugin........................			
" methylene-proto-catechuic, see Acid, piperonylic......................			
" **molybdic** (molybdenic, molybdænic), **chem. pure,**—free fr. Ammonium, Chlorine, Nitric acid;—[100% of MoO_3]....................		oz. .35	
" " **pure**........................		oz. .25	
" mono-brom-acetic		oz. 1.50	
" mono-chlor-acetic..................		oz. .50	

☞ When ordering, specify: "**MERCK'S**"!

	Containers incl.			
Acid, mucic (saccharo-lactic), pure	oz. .75			
" muriatic, see Acid, hydrochloric				
" niobic	15 gr. 1.00			
" nitric, crude,—sp. gr. 1.32 [50% NH O$_3$]				
" " ch. pure, " 1.185[30% "]; conform. to Ph. G. II	lb. .37			
" " " " " 1.20 [32% NH O$_3$]	lb. .38			
" " " " " 1.30 [48% "]	lb. .39			
" " " " " 1.40 [65% "]	lb. .40			
" " " " " 1.42 [69% "]; conform. to U. S. Ph. and Ph. Brit.	lb. .40			
" " fuming, (Nitroso-nitric acid), ch. pure,—sp. gr. 1.525	lb. .60			
" " " pure, according to Ph. G. II, —sp. gr. 1.48	lb. .65			
" nitro-hydrochloric (nitro-muriatic; chloro-nitrous, chlor-azotic),—[Aqua regia],—U. S. Ph.:—Mix 4 parts, by weight, of Nitric acid sp. gr. 1.42, and 15 of Hydrochloric acid sp. gr. 1.16				
" nitro-picric (nitro-phenisic, nitro-xanthic), see Acid, picric				
" œnanthic (œnanthic)	15 gr. .30			
" oleic (oleinic; elaic, elainic;—*not* elaidic, elaidinic, — which *see also!*), [Olein],—chem. pure,—U.S.Ph.	oz. 1.00			
" " commercial, clear	lb. .45			
" opianic (di-methyl-nor-opianic)	15 gr. 1.00			
" **ortho-phenol-sulphonic,** — in 33½-% solution,—see **Aseptol**				
" ortho-oxy-benzoic, see **Acid, salicylic**				
" osmic, so-called, see Acid, per-osmic, anhydrous				
" oxalic	lb. .35			
" " chem. pure	lb. .80			
" **oxalic, chem. pure, cryst.,—for analyses,**—[C$_2$H$_2$O$_4$. 2 H$_2$O.] — Large, colorless prisms; perfectly clearly soluble in Water; volatilizable without residue; free from Calcium, Iron, Sulphuric acid.—(*Oxalic acid of this degree of purity has never been in commerce hitherto,—having now first been introduced by me.*)	oz. .35			
" oxy-naphthoic, Alpha-.—(Reported as possessing 5-fold the anti-zymotic force of Salicylic acid;—also, as a good disinfectant.)	lb. 1.50			
" oxy-phenic (pyro-catechuic), see Pyrocatechin				
" oxy-succinic, see Acid, malic				
" palmitic (palmitinic), crude	lb. .75			
" " pure	15 gr. .35			
" para-tartaric, see Acid, uvic				
" parabanic	oz. 2.50			
" pectic (pectinic)	oz. 2.00			
" pelargonic,—from Oil of Rue (Ruta graveolens)				
" per-chloric, pure	oz. .50			
" per-osmic, anhydrous, (so-called "Osmic acid"), [Osmium Tetr-oxide]	15 gr. 2.00			
" phenic (phenylic), see Acid, carbolic				
" phenol-sulphonic (phenyl-sulphuric), see Acid, sulpho-carbolic				
" phenyl-glycollic, see Acid, mandelic				
" phloretic (phloretinic), see Phloretin				
" phospho-antimonic, — acc. to Otto, — Reagent for Alkaloids	oz. .35			
" " -molybdic,—solution [10%]	oz. .25			

☞ **When ordering, specify: "MERCK'S"!**

		Containers incl.	
Acid, phospho-wolframic (phospho-tungstic), cryst............................		oz.	.40
" " —solution [10%]..............		oz.	.30
" phosphoric, glacial (mono-hydric), [Metaphosphoric acid — HPO_3], in small lumps...................		lb.	.78
" " do., in sticks...................		lb.	.80
" " " chem. pure, cryst...........		lb.	1.00
" " officinal (tri-hydric), [Ortho-phosphoric acid — H_3PO_4], chem. pure,—sp. gr. 1.70, [85%],— syrupy consistency............		lb.	.65
" " do., liquid, chem. pure,—sp. gr. 1.12, [20% H_3PO_4], — Ph. G. II.............		lb.	.50
" " " " ch. pure, -sp. gr. 1.13, [22%]		lb.	.50
" " " " " " " 1.16, [27%]		lb.	.50
" " " " " " " 1.20, [32%]		lb.	.50
" " " " " " " 1.30, [45.5%]		lb.	.55
" " " " " " " 1.347, [50%], —.U. S. Ph.		lb.	.55
" " anhydrous, perfectly white, (Phosphoric Anhydride; Phósphorus Pent-oxide—P_2O_5)............		lb.	2.50
" phosphórous,—sp. gr. 1.12..........		oz.	.35
" phtalic, anhydrous, cryst. ⎫	(*Ortho*-phtalic acid)	oz.	.35
" " pure, cryst. .... ⎬		oz.	.50
" " crude............ ⎭		oz.	.25
" picr-amic (picr-aminic), cryst.........		oz.	1.00
" picric (picrinic, picro-nitric, nitro-picric, nitro-phenisic, nitro-xanthic; carb-azotic), cryst., pure......		oz.	.25
" " cryst., chem. pure..............		oz.	.30
" piperic (piperinic).................		oz.	2.50
" piperonylic (methylene-proto-catechuic)		15 gr.	.50
" plumbic, anhydrous, see Lead, peroxide........................			
" polygalic, (Polygalin), see Senegin....			
" propionic, pure......................		. oz.	1.50
" prussic, see Acid, hydrocyanic........			
" pyro-catechuic, see Pyro-catechin.....			
" pyro-gallic, subl., white ⎫	(Pyro-gallol)	oz.	.35
" " resubl.,—Ph. G. II ⎬		oz.	.39
" pyro-ligneous, purified, (Rectified Woodvinegar), [Acetum pyrolignosum rectificatum],—conforming to Ph. G. II.		lb.	.40
" pyro-tartaric, cryst...................		15 gr.	.35
" **quillayic** (quillayinic, quillayaic)........		15 gr.	2.00
" quinic (chinic, kinic), cryst............		oz.	3.00
" quino-picric (chino-picric, kino-picric).		oz.	4.00
" quinovic (chinovic, kinovic)..........		oz.	2.00
" racemic, see Acid, uvic..............			
" rheic (chrysophanic, *true*), see Rhubarb constituents: Rhein................			
" rosolic, (Ros-aurin)....................		oz.	.35
" rufigallic.........................		15 gr.	.25
" rutic, see Acid, capric................			
" saccharo-lactic, see Acid, mucic........			
" **salicylic**, (ortho-Oxy-benzoic acid), artificial, pure, amorphous........		lb.	1.90
" " **artificial, pure, cryst.**,—*U. S. Ph*...		lb.	2.00
" " " " re-crystalliz'd (dialyzed)		lb.	3.00
" " **natural,** from Oil of Wintergreen, (Oleum Gaultheriæ).........		oz.	.75
" salicylous, (ortho - Oxy - benz - aldehyd; Salicylic Aldehyd; Salicylal, Salicylol; Salicyl Hydride), — *true*, — [Essential Oil of Spiræa ulmaria]		oz.	5.00
" do., (do., etc.),—*synthetic*............		oz.	3.00
" santalic (santalinic), see Santalin......			

☞ **When ordering, specify: "MERCK'S"!**

	Containers incl.			
Acid, santoninic (*not* santonic!), cryst.,— [C₁₅H₂₀O₄].—(*Not* Santonin!)..				
" " anhydrous, [Santoninic Anhydride], see Santonin..........				
" **sclerotic** (sclerotinic), acc. to *Dragendorff*.	15 gr. .25			
" " according to Podwyssotzki.......	15 gr. .35			
N.B.— *See, also:* **Acid, ergotic.**				
" **scoparic,** see **Scoparin**...............				
" sebacylic, cryst.................	oz. 1.25			
" selenic, pure, (Selenic Hydroxide),—sp. gr. 1.40........	oz. 4.00			
" selenious, anhydrous, sublimed, (Selenious Oxide)............	oz. 5.00			
" silicic, (Silicic Oxide), [Silica, Silicea; Silex], pure, natural, pulverized	lb. .80			
" " pure, by wet process; dried.....	lb. 1.25			
" silvic (silvinic)	lb. 3.50			
" sorbic (sorbinic), cryst......	15 gr. .50			
" **sozolic** (ortho-phenol-sulphonic,—in 33⅓-% solution),—see **Aseptol**........				
" stannic, anhydrous, see **Tin,** oxide, white				
" stearic (stearinic), pure............	oz. 1.50			
" stibic, anhydrous, see **Antimony,** oxide, white, *true*, (Pent-oxide)...........				
" stibious, anhydrous, see **Antimony,** oxide, precipitated, pure, (Tri-oxide)...				
" suberic	15 gr. .50			
" succinic, crude, sublimed ...) (Volatile	lb. 1.00			
" " purified,—Ph. G. I.... } Salt of	lb. 1.50			
" " pure,—perfect. colorless) Amber)	oz. .22			
" sulpho-anilic (sulph-anilic), cryst., white	oz. .50			
" sulpho-carbolic (sulpho-phenylic, sulpho-phenic; phenol-sulphonic, phenyl-sulphuric),—[Sulpho-phenol, Sulpho-carbol],—containing *both* the "Para-" and the "Ortho-" acid............	oz. .25			
" " **Ortho-,** pure,—in 33⅓-% aqueous solution,—see **Aseptol**				
" **sulpho-ichthyolic,** see under **Ichthyol preparations**.................				
" sulpho-naphthyl-aminic	oz. .40			
" sulpho-phenic (sulpho-phenylic), see **Acid, sulpho-carbolic**...........				
" sulpho-vinous (ethyl-sulphurous),—sp. gr. 1.1;—[*not* identical with: Sulphovinic (Ethyl-sulphuric) acid!]....	oz. .30			
" sulphuric, ch. pure,—sp. gr. 1.840, [97% H₂SO₄],—*U. S. Ph.,*—(Monohydrated Tri-oxide of Sulphur)	lb. .40			
" " crude,—free from Arsenic,—(so-called "Oil of Vitriol"),—[66° Bé]				
" " anhydrous, pure, (Sulphuric Anhydride; Tri-oxide of Sulphur)	100 grammes : 1.00			
" " " commercial				
" sulphurous, (Hydrated Sulphurous Oxide [Di-oxide]),—solution; sp. gr. 1.022-1.026, [about 5-6% of SO₂]..............	lb. .40			
" " —do.,—[3.5%],—*U. S. Ph.*......	lb. .30			
" " glycerolate (glycerite) of, [solution in Glycerin], see **Glycerin, sulphurous**..............				
" **tannic,** see **Tannin**.................				
" tantalic, (Hydrated Tantalic Oxide [Pent-oxide]);—white powder,—prepared from Tantalic Chloride...........	15 gr. 2.00			

☞ **When ordering, specify : "MERCK'S" !**

MERCK'S INDEX. 9

		Containers incl.			
Acid, tartaric, Dextro-, — (Essential Salt of Tartar, — *not to be confounded with:* "Salt of Tartar" = Pure Potassium Carbonate from the Bi-tartrate!), —pure, cryst.		lb. .90			
" " do., pure, powder		lb. .90			
" " " chem. pure, cryst., — conform. to the requirements of *U. S. Ph.* and the other Pharmacopœias		lb. 1.25			
" " " chem. pure, powder		lb. 1.25			
" " Para-, see Acid, uvic					
" tartronic		15 gr. 1.50			
" tauro-cholic (choleic, choleinic)		15 gr. 2.00			
" telluric, di-hydrated, (Tri-hydrated Telluric Oxide [Tri-oxide]; Di-hydrated Telluric Hydroxide)		15 gr. 1.50			
" tellurous, (Hydrated Tellurous Oxide [Di-oxide]; Tellurous Hydroxide)		15 gr. 1.40			
" terpenylic (turpenylic), dry		15 gr. .75			
" thio-phosphorous, anhydrous, see Phosphorus, tri-sulphide					
" thymic, (Thyme-camphor), see Thymol					
" tiglic (tiglinic), see Acid, crotonolic					
" titanic, Ortho-, (Titanic Hydroxide; Di-hydrated Di-oxide of Titanium)		oz. 1.50			
" tri-chlor-acetic		oz. .50			
" tri-chlor-methyl-sulphonic, see Tri-chlor-methyl, sulphite					
" tri-cyanic, see Acid, cyan-uric					
" tropic		15 gr. 1.00			
" tungstic, anhydrous, see Acid, wolframic, anhydrous					
" turpenylic, see Acid, terpenylic					
" uranic, anhydrous, see Uranium, oxide, red					
" ureous, (Uric Oxide), see Xanthine					
" uric (lithic), pure		oz. .80			
" uvic (para-tartaric; racemic)		oz. 1.00			
" valerianic (valeric; amylic), [the so-called Tri-hydrate], —Ph. G. I.	All Iso-valeric acids.	oz. .35			
" " pure, (the so-called Mono-hydrate),—formerly officinal		oz. .40			
" " from Valerian-root		oz. 1.00			
" vanadic (vanadinic), Meta-, [Hydrated Pent-oxide of Vanadium; Vanadic Hydroxide], chem. pure		oz. 8.00			
" " do., commercial		oz. 3.50			
" vanillic (vanillinic) [methyl-proto-catechuic]		15 gr. .50			
" veratric (di-methyl-proto-catechuic), cryst.		15 gr. 1.00			
" vieiric, see Vieirin					
" wolframic (tungstic), anhydrous, [Tungstic (Wolframic) Oxide (Tri-oxide)], crude		lb. 2.00			
" " do., pure		oz. .40			
Aconitine Merck (Aconitia),–from Aconitum napellus Linné, [sometimes called Napellus Stoerckeanum]:					
pure, amorphous, powder		⅛ oz. vls. oz. 11.00			
" cryst.		15 gr. 2.00			
arseniate (arsenate)		15 gr. 1.00			
hydrobromate		15 gr. 1.00			
hydrochlorate		15 gr. 1.00			
nitrate, amorphous		15 gr. 1.00			
" cryst.		15 gr. 1.75			
oleate, [66⅔% of Aconitine]		15 gr. 2.00			

☞ **When ordering, specify: "MERCK'S"!**

	Containers incl.			
Aconitine Merck (Aconitia),--*continued*:				
salicylate, cryst..........................	15 gr. 1.00			
sulphate................................	15 gr. 1.00			
valerianate.............................	15 gr. 1.00			
Aconitine from Aconitum ferox, (Bish or Bikh root; Nepaul Aconite),—[the so-called British Aconitine—Aconitinum anglicum—*Pseudo-Aconitine*]............	15 gr. 2.50			
" from Japanese Aconite-root.............	15 gr. 1.25			
Acorn-sugar, see Quercit..................				
Adonidin.................................	15 gr. 3.00			
" tannate..............................	15 gr. 3.00			
Ærugo purificata; and, do. destillata;—see Copper, acetate, basic; and, normal, *U. S. Ph.*				
Æsculin, Æthal, Æther, etc., **Æthiops,** etc.; see Esculin, Ethal, Ether, etc., Ethiops, etc...............................				
Æth-oxy-Caffeine, see Ethyl-oxy-Caffeine				
Æthyl, Æthyl-amine, Æthylene, Æthylidene,—etc.;—see Ethyl, Ethyl-amine, Ethylene, Ethylidene,—etc...............				
Agaricin Merck, chem. pure,—from White Agaric, (Fungus laricis);—free from purgative resin.—[*Not* identical with Laricic (Agaricic) Acid,—which *see also!*]..................	15 gr. .25			
Alant-camphor, solid, see Helenin............				
" **liquid,** see Alantol..................				
Alant-starch (*Alantin*), see Inulin.........				
Alantol (*not* Alantin!).—[The *liquid* Alant-, or Elecampane-, or Inula-camphor.]—(An internal antiseptic.)....................	⅛ oz. vls. oz. 20.00			
N. B.—*Compare, also:* **Helenin.**				
Albumen, Egg, (Albumen ovi), dried, see under **Egg preparations**....................				
N. B.—*See, also:* Yelk, dried,—under **Egg preparations.**				
Albumin,—*from eggs,*—soluble............	lb. .85			
" *fr. eggs,* I,—soluble,—inodorous;—its aqu. solution is of sp. gr. 1.03	lb. 1.50			
" " " soluble,—in scales;—absolutely free from Fibrinous matter;—for laboratory use..........				
" " " soluble,—impalpable powder;—for gilders', stampers' and bookbinders' uses...........				
" *from blood.*.........................	lb. .50			
" " " chem. pure....................	oz. .65			
" iodized, see Iodine, albuminated.......				
Albumin, Iron-, in scales; and do., peptonized; and do., saccharated;—see Iron, albuminate, etc.; etc.; etc.................				
N. B.—*Compare, also:*				
Iron, lactate ....⎱				
" phosphate........ ⎬ *albuminated.*				
" pyro-phosphate... ⎰				
—*Other Metallic Albuminates,* see likewise under the respective metals.				
Alcohol (Ethylic alcohol), "absolute"— I, —sp. gr. 0.796, [about 99%].........	lb. 1.50			
" (Ethylic alcohol), "absolute"— II,—sp. gr. 0.805-0.808, [about 95-97%]..	lb. 1.45			
" (Ethylic alcohol),—*U. S. Ph.,*—sp. gr. 0.820, [about 91%]................	lb. 1.25			
" allylic.............................	lb. 10.00			
" ammoniated, see Ammonia, Spirit of ..				
" amylic, primary, (Iso-pentylic alcohol; Iso-butyl-carbinol), [so-called "Fusel-oil"]......	lb. .40			
" " " pure,—boiling-point 128-130° C [262.4-266 F].....	lb. .60			

☞ **When ordering, specify : "MERCK'S"!**

	Containers incl.		
Alcohol, amylic, primary, — (*as above!*);— chem. pure...	lb. .75		
" amylic, tertiary, see **Amylene Hydrate**....			
" benzylic...	oz. 2.50		
" " ortho-Oxy-, see Saligenin...			
" butylic, Iso-, (Iso-propyl-carbinol),— b.-pt. 107–110° C [224.6–230 F]	lb. 2.00		
" " tertiary, see Tri-methyl-carbinol.			
" caprylic...	oz. 1.00		
" caustic, see Sodium, ethylate, cryst....			
" cetylic, (Cetyl-alcohol), see Ethal...			
" cinnamic (cinnamylic; styrylic), [Cinnam-alcohol; Styrol-alc.], see Styrone			
" ethylenic, see Ethylene-glycol...			
" hydrochlorated, see Spirit of Muriatic Ether...			
" iso-butylic, see Alcohol, butylic, Iso-...			
" iso-pentylic, see Alcohol, amylic, primary...			
" iso-propylic, see Alcohol, propylic, Iso-			
" methylic, (Wood-spirit, Wood-naphtha; Wood-alcohol; Pyro-ligneous [pyro-xylic] Spirit; Carbinol, Methol),—pure...	lb. 1.00		
" " chem. pure,—b. p. 64–70° C [147–158 F]...	lb. 1.25		
" " [94–95%]...	lb. 1.00		
" " [90%]...	lb. .50		
" ortho-oxy-benzylic (salicylous), see Saligenin...			
" propylic, (Ethyl-carbinol), — b.-pt. 96–99° C [204.8–210.2 F]...	lb. 6.00		
" " Iso-, (Di-methyl-carbinol)...	oz. 2.00		
" salicylous (ortho-oxy-benzylic), see Saligenin...			
" styrylic, (Styrol-alcohol), see Styrone..			
" —*so-called*—of Sulphur, ("Alcohol Sulphuris"), see Carbon, bi-sulphide...			
" Thio-, ethylic, see Mercaptan, ethylic..			
" Wood-, see Alcohol, methylic...			
Aldehyd (Acetic [Ethylic] aldehyd), commercial...	lb. 1.00		
" concentrated...	lb. 1.25		
" highly concentrated...	lb. 2.50		
" absolute...	lb. 6.00		
Aldehyd, Iso-butyl-, see Iso-butyl-aldehyd			
" salicylic, (ortho-Oxy-benz-aldehyd), see Acid, salicylous...			
Aldehyd-Ammonia (Ammoniated Acetic [Ethylic] Aldehyd), pure, cryst...	oz. .85		
Algaroth, Powder of, see Antimony, oxychloride...			
Alizarin, paste...	lb. 1.00		
Alkannin (Anchusin), inspissated Extract of	oz. .75		
" insp., wholly soluble in Alcohol Alkanet.	oz. 1.00		
Alkargen (*not* Alkarsin!), see Acid, cacodylic			
Allantoin...	15 gr. .50		
Alloxan...	15 gr. .25		
Alloxantin...	15 gr. .35		
Allyl, bromide (mono-bromide)...	oz. 2.00		
" iodide...	oz. 2.25		
" sulpho-cyanate (thio-cyanate), — synthetical;—see Essential Oils: Mustard, Black,—artificial...			
" tri-bromide...	oz. 2.00		
Allyl-amine...	15 gr. .50		
Aloe Purple...	oz. 2.00		
Aloin (Barb-aloin), chem. pure...	oz. .30		
Alstonine, see Chlorogenine.			
Althein (Altheine), see Asparagin...			

☞ When ordering, specify: "**MERCK'S**"!

	Containers incl.		
Alum, ammoniacal, (Ammonium-alum, Ammonia-alum), [Aluminium and Ammonium, sulphate]	lb. .35		
" " pure,—*Alumen,* Ph. Brit.	lb. .40		
" ammonio-ferric, (Ammoniacal Iron-alum), see Iron, Sesqui-compounds: Ammonio-ferric sulphate			
" caesic (cæsic), [Caesium-alum]	15 gr. 1.00		
" chromic, (Chrome-alum), [Chromium and Potassium, sulphate], large cryst.	lb. .40		
" " II	lb. .35		
" Copper-, so-called,—("Divine Stone"),—see Copper, aluminated			
" ferric, (Iron-alum), [Aluminium and Iron, sulphate; Aluminio-ferric Sulphate]	lb. .40		
" potassic, (Potassium-alum, Potassa-alum', [Aluminium and Potassium, sulphate], chem. pure, cryst.	lb. .50		
" " chem. pure, powder	lb. .55		
" " " " impalpable powder	lb. .60		
" " Ph. G. II., cryst.,—*Alumen, U. S. Ph.*	lb. .40		
" " " " powder	lb. .45		
" " free from Iron	lb. .35		
" " —caustic pencils, turned,—with or without wooden casing	doz. 1.00		
" " crude, large crystals	lb. .25		
" " burnt (dried, exsiccated), lumps *Alumen exsiccatum, U. S. Ph.*	lb. .30		
" " " powder	lb. .35		
" potassio-ferric, (Potassic Iron-alum), see Iron, Sesqui-compounds: Potassio-ferric sulphate			
" rubidic, (Rubidium-alum)	15 gr. .50		
" sodic, (Sodium-alum, Soda-alum'), [Aluminium and Sodium, sulphate], commercial, cryst.	lb. .50		
" " pure	lb. .65		
" zincic, (Zinc-alum), [Aluminium and Zinc, sulphate]	lb. 1.00		
" " in sticks	lb. 1.50		
Alumina (Argilla pura—Pure Argil), anhydrous, chem. pure, see Aluminium, oxide, anhydrous			
" hydrated,—commercial; and: pure, *U.S. Ph.;*—see Aluminium, oxide, precipitated, etc.; etc.			
Alumina Purple of **Gold,** see Gold, Alumina Purple of			
Aluminated Copper, (so-called "Copperalum"), see Copper, aluminated			
Aluminium (Aluminum), double salts of, see "Aluminium and —" (below!)			
" metallic, bar	oz. 1.25		
" " sheet	oz. 2.00		
" " " thin	oz. 1.75		
" " wire	oz. 2.00		
" " powder, coarse	oz. 2.00		
" " " impalpable	oz. 2.50		
" " leaf,—book of 250 leaves	Book 2.00		
" acetate, pure, liquid,—[5% of Basic Aluminium acetate]	lb. .40		
" " " " Ph. G. II.,—[8% do.]	lb. .50		
" " " " dry	lb. 1.50		
" aceto-glycerolate, (Glycerolate [Glycerite] of Acetate of Aluminium)	oz. .30		
" " -tartrate, dry	oz. .25		

☞ **When ordering, specify: "MERCK'S"!**

	Containers incl.		
Aluminium, arseniate (arsenate)	oz. .30		
" benzoate	oz. 1.50		
" bromide	oz. .50		
" chloride, pure, dry	lb. 1.25		
" " II	lb. 1.20		
" cinnamate, pure, cryst.			
" fluoride	oz. .40		
" hydrate (hydroxide), *U. S. Ph.* and Ph. G. I, —see Aluminium, oxide, precipitated, *pure.—[Argil,* see same, *con'l.*].			
" nitrate, pure	lb. 2.00		
" " II	lb. 1.50		
" " solution [15° Baumé]	lb. 1.25		
" oleate	oz. .30		
" oxalate, pure	oz. .30		
" oxide, anhydrous, (Anhydrous Alumina), chem. pure,—[Argilla anhydrica purissima]	oz. .50		
" " precipitated (hydrated), commercial, [Argil]	lb. .40		
" " " pure, (Hydrate [Hydroxide] of Aluminium),—*U.S.Ph.;* —[Hydrated Alumina, — Argilla hydrata pura, Ph. G. I]	lb. 1.10		
" palmitate, pure	lb. 1.50		
" " crude	lb. 1.10		
" phosphate	oz. .40		
" rhodanide, see Alumin., sulpho-cyanate.			
" silicate	oz. .25		
" sulphate, twice refined,—free from Iron	lb. .25		
" " pure,—*U. S. Ph.* and Ph. G. II	lb. .75		
" " chem. pure, cryst.	lb. 1.25		
" sulpho-carbolate (phenol-sulphonate, sulpho-phenate)	oz. .50		
" sulpho-cyanate (thio-cyanate; rhodanide)	oz. .50		
" " —solution [20° Baumé]	lb. 1.00		
" tannate	oz. .40		
" tartrate	oz. .25		
" " pure	oz. .40		
" thio-cyanate, see Al., sulpho-cyanate			
Aluminium and Ammonium, sulphate, see Alum, ammoniacal			
" and **Iron,** sulphate, see Alum, ferric			
" and **Potassium,** sulphate, see Alum, potassic			
" and **Sodium,** chloride, cryst.	oz. .30		
" " " sulphate, see Alum, sodic			
" and **Zinc,** sulphate, see Alum, zincic			
N. B.—Other *Double*—(also *Triple*)—*Sulphates,* see likewise under Alum.			
Amalgams: of Sodium; of Zinc; and, of Zinc and Tin;—see under the respective *metals*			
Amidin, iodized, see Starch, iodized			
Amido-benzene (-benzol), see Aniline			
Amido-ethane, see Ethyl-amine			
Amido-methane, chloride, see Methyl-amine, chloride			
Amido-phenol (Ox-aniline), ortho-, hydrochlorate	15 gr. .75		
Amido-toluene (-toluol), see Toluidine			
Amido-xylene (-xylol), see Xylidine			
Ammon, see Ammonium			
Ammonia, Spirit (Alcoholic Solution) of,—acc. to Dzondi, — [Liquor Ammonii caustici spirituosus], (Ammoniated Alcohol),—sp. gr. 0.810	lb. .85		
" Spirit of, aromatic, see Spirit of Ammonia, aromatic			

☞ When ordering, specify: "**MERCK'S**"!

		Containers incl.		
Ammonia, **Water** (Aqueous Solution) of, [Aqua Ammoniæ, Liquor Ammoniæ, L. Ammonii caustici], pure, —sp. gr. 0.875, [abt. 40% N H$_3$]		lb. .60		
" do. do., pure, -sp. gr. 0.885, [" 36% "]		lb. .55		
" " " " " 0.890, [" 33% "]		lb. .50		
" " " " " 0.900, [" 29% "], —*Aq. Amm. fortior, U. S. Ph.*		lb. .40		
" " " " sp. gr. 0.910, [abt. 25% N H$_3$]		lb. .35		
" " " " " 0.925, [" 20% "]		lb. .30		
" " " " " 0.960, [" 10% "], —Ph. G. II.; —*Aq. Amm., U. S. Ph.*		lb. .25		
" " " technically pure, —various grades				
Ammonia Alum, see Alum, ammoniacal..				
Ammoniacal Iron-Tartar, see Iron, Sesquicompounds: Ammonio-Ferric tartrate, *U. S. Ph.*				
" **Turpeth,** see Mercury and Ammonium, sulphate				
Ammoniated Alcohol, see Ammonia, Spirit of....................				
" **Aldehyd,** see Aldehyd-Ammonia....				
" **Copper,**—so-called,—see Copper and Ammonium, sulphate				
" **Glycyrrhizin,** *U. S. Ph.*, see Gl., ammoniated				
" **Iron,** —so-called, —see Ammonium, chloride, with Ferric Chloride.......				
" **Mercury,**—so-called:—infusible (*U. S. Ph.*); and, fusible;—see Mercury, ammoniated, etc.; etc...........				
" **Tartar, soluble,** see Potassium and Ammonium, tartrate				
Ammonio- double and triple salts, see "Ammonium and —" (below!)				
Ammonium (Ammon), acetate, cryst......		oz. .30		
" acetate, solution, (so-called "Spirit" of Mindererus), see under Solutions....				
" arseniate (arsenate), cryst............		oz. .30		
" arsenite		oz. .30		
" benzoate, from true Benzoic Acid prepared from Benzoin-resin.....		oz. .40		
" " —*U. S. Ph.*,—fr. artificial do. do.		oz. .30		
" bi-carbonate, cryst....................		oz. .30		
" **bi-chromate, cryst., chem. pure,**—free fr. Sulphate of Potassium		lb. 1.25		
" bi-malate, cryst.		oz. 2.00		
" bi-oxalate (bin-oxalate), chem. pure....		oz. .30		
" " commercial		oz. .25		
" bi-phosphate		oz. .25		
" bi-sulphate		oz. .30		
" bi-sulphite		oz. .50		
" bi-tartrate....................		oz. .40		
" borate....................		oz. .30		
" " pure....................		oz. .45		
" boro-citrate		oz. .50		
" bromide, conform. to *U. S. Ph.* & Ph. G. II		lb. .90		
" camphorate		oz. 3.00		
" carb-amate (carb-aminate), [so-called Anhydride of Ammonium Carbonate].—Exceedingly volatile..........		oz. 1.50		
" carbolate, see Ammonium, phenate....				
" carbonate....................		lb. .50		
" " chem. pure,—*U. S. Ph.*.........		lb. .60		
" " anhydrous,—so-called,—see Ammonium, carb-amate............				
" chlorate, per-, see Ammon., per-chlorate				
" chloride, (Sal ammoniacum), semi-purif.				
" " purified, white		lb. .28		
" " chem. pure,—*U. S. Ph.* & Ph. G. II.		lb. .40		
" " sublimed, in lumps		lb. .50		

☞ When ordering, specify: "**MERCK'S**"!

		Containers incl.		
Ammonium, chloride, with Ferric Chloride,–(Ammonio-chloride of Iron; so-called "Ammoniated Iron"),—Ph. G. II		lb. .60		
Ammonium, chloro-stannate, (Ammonio-stannic Chloride), [Pink (Dyers') Salt], see Tin and Ammonium, chloride				
" chromate, neutral, pure		oz. .50		
" citrate		oz. .25		
" Cuprico-, double salts of, see under Copper and Ammonium				
" fluoride		oz. .40		
" formate, pure		oz. 1.00		
" gallate, neutral		oz. 1.25		
" **glycyrrhizate,** pharmacopeial, see **Glycyrrhizin, ammoniated, soluble,**—*U. S. Ph.*				
" hydro-sulphuretted solution of sulphide, (*Hydrothion-ammonium* Solution), — see Solutions: Ammonium sulphide,—hydro-sulphuretted				
" hypo-phosphite		oz. .50		
" hypo-sulphite, see Ammonium, thio-sulphate				
" **ichthyol-sulphonate** (sulpho-ichthyolate), [Ichthyol], see **Ichthyol preparations,** etc.				
" iodide,—*U. S. Ph.* and *Ph. G. II.*		oz. .45		
" lactate		oz. .50		
" mellitate (mellate), cryst.		oz. 5.00		
" **molybdate** (molybdenate), **chem. pure**		oz. .45		
" nitrate		lb. .40		
" " pure; cryst.		lb. .45		
" " " dry		lb. .60		
" " " fused		lb. .65		
" " chem. pure, cryst.,—*U. S. Ph.*		lb. .60		
" nitrite, liquid		oz. .30		
" oxalate, (Di-ammonium oxalate), pure		lb. .90		
" " (do.), chem. pure		lb. 1.00		
" oxal-urate (uro-oxalate)		oz. .50		
" per-chlorate		oz. 2.00		
" phenate (phenylate, carbolate)		oz. .25		
" phosphate, (Di-ammonium ortho-Phosphate), purified, cryst.		lb. .75		
" " (do.), pure		lb. 1.00		
" " " chem. pure,—*U. S. Ph.* and *Ph. G. I.*		lb. 1.10		
" phosphite		oz. .50		
" phospho-molybdate		oz. 1.25		
" picramate		oz. 3.00		
" picrate (picro-nitrate)		oz. .35		
" picro-carminate, dry		oz. 1.50		
" purpurate, see Murexid				
" rhodanide, see Ammon., sulpho-cyanate.				
" salicylate, cryst.		oz. .50		
" seleniate (selenate)		oz. 6.00		
" succinate, pure, cryst.		oz. .35		
" sulphate, crude		lb. .30		
" " pure		lb. .39		
" " chem. pure,—*U. S. Ph.*		lb. .50		
" sulphide (sulphuret),-hydro-sulphuretted solution of;—see Solutions: Ammon. sulphide,—hydro-sulphuretted.				
" sulphite		oz. .25		
" sulpho-carbolate (phenol-sulphonate, sulpho-phenate)		oz. .30		
" sulpho-cyanate (thio-cyanate; rhodanide), pure		lb. 1.00		
" " commercial		lb. .75		
" **sulpho-ichthyolate** (ichthyol-sulphonate), [Ichthyol], see **Ichthyol preparations,** etc.				
" tannate, liquid		oz. .30		
" tartrate, neutral, cryst.		oz. .25		

☞ When ordering, specify : "MERCK'S" !

	Containers incl.			
Ammonium, thio-cyanate, see Ammonium, sulpho-cyanate................				
" thion-urate..................	oz. 2.00			
" thio-sulphate (formerly called "hyposulphite"), pure..............	oz. .30			
" tungstate, see Ammonium, wolframate.				
" uranate, (so-called "Hydrated Oxide of Uranium"), [also sometimes called "Uranium Yellow,"—which latter name properly applies to *Sodium* Uranate]...................	oz. 1.00			
" urate, pure.................	oz. .50			
" uro-oxalate, see Ammonium, oxal-urate				
" valerianate, cryst., white,—*U. S. Ph*...	oz. .32			
" vanadate, chem. pure.............	oz. 2.00			
" wolframate (tungstate)............	oz. .40			
Ammonium and **Aluminium,** sulphate, see Alum, ammoniacal.............				
" and **Bismuth,** citrate, see Bismuth and Ammonium, citrate, *U. S. Ph*.......				
" and **Cadmium,** salts, see Cadm. & Amm.				
" and **Cobalt,** sulphate, see C. & A., sulph.				
" and **Copper,** salts, see Copper and Am.				
" and **Iron,** arsenicico-citrate, see Iron, arseniate and citrate, ammoniated..............				
" " " chloride, (so-called "Ammoniated Iron"), see Ammonium, chloride, w. Ferric Chloride				
" " " divers salts, see Iron, Monocompounds; and Iron, Sesqui-compounds, —(*the latter embracing the U.S.Ph. salts:* Citrate; Sulphate; Tartrate)				
" and **Magnesium,** salts, see Magn. & A.				
" and **Mercury,** salts, see Merc. & Amm.				
" and **Nickel,** salts, see N. & Ammonium				
" and **Platinum,** double *and triple* salts, see Platinum double Chlorides; do. double Cyanides; do. *triple* Cyanides; and do., divers double Salts.........				
" and **Potassium,** salts, see Pot. & Amm.				
" and **Silver,** salts, see Silver and Ammon.				
" and **Sodium,** salts, see Sodium and A.				
" and **Tin,** chloride, (Pink Salt; Dyers' Salt), see Tin and Ammon., chloride.				
" and **Zinc,** chloride, see Z. & A., chloride				
Ammonium, Platinum, and : } Calcium, cyanuret........ } See under Platinum triple Cyanides. Copper, cyanuret-cyanide . }				
Ammonium, Solutions of *divers salts* of, see under Solutions.................				
Amygdalin...................	⅛ oz. vls. oz. 2.00			
Amyl ("Amylium"—*not* Amylum!), acetate, [Amylo-acetic Ether], (so-called "Pear-oil")...................	lb. 4.00			
" do., [etc.], (etc.),—chem. pure.......	lb. 4.50			
" bromide...................	oz. .50			
" butyrate..................	lb. 5.00			
" chloride..................	oz. .60			
" cyanide, (Cyano-amyl), [Capro-nitrile]...				
" formate...................	oz. .50			
" hydride, (Pentane), crude, see Eupione				
" iodide,—b.-pt. 140–148° C [284–298.4 F]	oz. .80			
" nitrate...................	oz. .50			
" nitrite, (Amylo-nitrous Ether)........	oz. .29			
" " in lymph-tubes of 1–3 drops.....				
" " pure,—*U. S. Ph.* and Ph. G. II...	oz. .30			
" oxide, hydrated, (so-called "Fusel-oil"), see Alcohol, amylic.............				

☞ When ordering, specify: "**MERCK'S**"!

	Containers incl.			
Amyl, phenate (carbolate), [Amyl-phenol], cryst.—(A hypnotic.)	oz. 2.50			
" valerianate, (so-called "Apple-oil")	lb. 6.00			
Amyl-phenol, see Amyl, phenate				
Amylene	oz. .50			
" bromide	oz. 1.00			
Amylene Hydrate, (Tertiary Amylic Alcohol),—boiling-point 100° C [212 F],—sp. gr. 0.81.—(An excellent hypnotic, not materially affecting the heart-action.)	oz. .75			
Amylum iodatum, (Iodized Starch), U. S. Ph., see Starch, iodized				
Amylum, animal, — so-called, — see Glycogen				
Analgesine, so-called, see Antipyrine				
Anchusin, see Alkannin				
Anemonin (Anemone-camphor, Pulsatilla-camphor)	15 gr. 1.75			
Anethol, liquid	oz. 1.00			
Anethol-Quinine, see Quinine, anisated				
Aniline (Anilia), [Amido-benzene (-benzol); Benzid-am; Phenyl-amine], pure	lb. 1.00			
" acetate	oz. .50			
" chloride	oz. .30			
" nitrate	oz. .30			
" oxalate	oz. .40			
" sulphate	oz. .30			
Aniline, di-Methyl-, see Di-methyl-aniline				
" **Methyl-,** see Methyl-aniline				
Aniline and Phenol Dyes (or *Colors*):				
Aurin	oz. .40			
Black, Nigrosine, soluble in Water	lb. 2.25			
" " " in Alcohol	lb. 2.50			
Blue,—free from Arsenic	oz. .75			
" permanent, — soluble in Alcohol; — free from Arsenic	oz. .65			
" Ethylene-	oz. .75			
" Methylene-	oz. .60			
" Naphthalene-	oz. .60			
" Phenyl-,—free from Arsenic	oz. .65			
" reddish	oz. .60			
Brown, Bismarck-	lb. 2.00			
" Vesuvine-	lb. 3.50			
Chrysoidine,—free from Arsenic	lb. 2.50			
Green, Malachite-, cryst.,—free fr. Arsenic	lb. 2.50			
" " powder, " " "	lb. 2.00			
" Methyl-,—free from Arsenic	lb. 2.50			
" Iodine-	oz. 2.00			
" brilliant	oz. .25			
Induline,—free from Arsenic	oz. .50			
Orange, Helianthine	oz. .75			
" Di-methyl-aniline-	oz. .65			
" Ethyl-	oz. .45			
" Methyl-,—free from Arsenic	oz. .50			
Phosphine, so-called, see Aniline and Phenol Dyes: Yellow, Chrys-aniline				
Purpurin: dry; and, paste,—see *Purpurin*				
Red, Fuchsine,-free fr. Arsenic;-large cryst.	oz. .40			
" Congo-	oz. .50			
" Corallin	oz. .40			
" Eosin	oz. .50			
" Magdala-	15 gr. 1.00			
" ruby S-	oz. .40			
" " orange-	oz. .35			
" Safranine	oz. .65			
" scarlet,—free from Arsenic	oz. .30			
Rose, Bengal-, " " "	oz. 1.00			
Tropeolin (Tropæolin), see *Tropeolin*				
Violet, Gentian-,—free from Arsenic	oz. .30			
" Methyl-,— " " "	oz. .25			

☞ **When ordering, specify: "MERCK'S"!**

		Containers incl.			
Aniline and Phenol Dyes (or *Colors*),—*continued*:					
Violet, Hoffmann's.......................		oz.	.40		
Yellow		oz.	.25		
" Chrys-aniline (sometimes also called Phosphine),—free from Arsenic.		oz.	.75		
" Luteoline		oz.	.25		
" Manchester-.......................		oz.	.25		
.. Martius-		oz.	.30		
" Naphthalene-		oz.	.40		
" Primrose- (Primula-)		oz.	.25		
:: Safranine					
.. T-.................................		oz.	.50		
" orange T-,—free from Arsenic.....		oz.	.25		
Aniline, Ros-, see Ros-aniline ....,..					
Anisol (Methyl Phenate; Methylo-phenic Ether)...................................		oz.	2.00		
Anthracene, purified, sublimed...........		oz.	.25		
Anthraco-potassa (Anthrako-kali), simple.		oz.	.25		
" sulphurated........................		oz.	.25		
Anthra-quinone (A.-chinone, A.-kinone)..		oz.	.50		
Anthrarobin (Anthro-arobin).— A derivative from Alizarin, etc.—[Used as a mild succedaneum for *Chrysarobin* (that is: the so-called "Medicinal Chrysophanic Acid").].		oz.	.50		
Antichlors (Anti-chlorines), see Sodium: thio-sulphate; bi-sulphite; and, sulphite..					
Antifebrin, perf. white, chem. pure, cryst., Kalle's, —under my conjugate guarantee for purity; —(*Medicinal* Phenyl-acet-amide, *Medicinal* Acet-anilide).—[Lately *very prominent* as an analgetic, anodyne, sedative, and hypnotic, —in hemicrania, neuralgias, dysmenorrhea, insomnia, delirium, etc.]		oz.	.25		
Antifungin		oz.	.35		
Antimonial Crocus (Saffron), see Potassa, antimonio-sulphurated, *washed*......					
" **Ethiops,** (Antimony and Mercury Black Sulphides), see Mercury, antimonio-sulphide....................					
" **Glass,** see Antimony, sulphide, vitreous,—so-called					
" **Powder,** *U. S. Ph.*,—(James's Febrile Powder), [Antimonious Oxide with Calcium Phosphate]................		lb.	1.50		
Antimoniated (Stibiated) **Liver of Lime,** [Stibiated Calcic Liver of Sulphur], see Lime, antimonio-sulphurated....					
" **Tartar,** (Tartar Emetic; Tartarated Antimony), see Antimony and Potassium, tartrate, *U. S. Ph.*; and other grades.					
Antimony (Antimonium; Stibium), double salts of, see "Antimony and—" (below!).............................					
" metallic...... } Regulus of Antimony		lb.	.35		
" " ch. pure. }		oz.	.25		
" arseniate (arsenate)..................		oz.	.30		
" arsenite		oz.	.30		
" bromide...........................		oz.	.50		
" chloride, Antimonious, (tri-chloride), pure, cryst.,—[*Concentrated* Butter of Antimony]...............		oz.	.30		
N.B.— *See, also:*—Solutions: Antimonious chloride, — (*Liquid* Butter of Antimony)					
" " Antimonic, (penta-chloride), see Antimony, per-chloride					
" diaphoretic, washed (purified), see Potassium, antimonate, *pharmacopeial*					
" " unwashed, see do., do., *crude*....					
" iodide, cryst.............................		oz.	1.00		

☞ **When ordering, specify : "MERCK'S"!**

	Containers incl.
Antimony, oxalate..........................	lb. 1.25
" oxide, white,—*true*,—Ph. Bor.V; (Antimonic oxide,—Pent-oxide),[Anhydrous Stibic or Antimonic Acid—Sb_2O_5].................	lb. .75
" " do.,—*so-called*,—Ph. Bor. VI;—(*Washed* [purified] *Diaphoretic* Antimony; Calx Antimonii [Stibii]),—[principally : $KSbO_3$], —see Potassium, antimonate, *pharmacopeial*.................	
" " diaphoretic, unwashed,—*so-called*, —(*Unwashed* Diaphoretic Antimony), see Potassium, antimonate, *crude*.................	
" " precipitated, (Antimonious oxide, —Tri-oxide), pure,—*Antimonii oxidum, U. S. Ph.;*—[Stibium oxydatum præcipitatum, Ph. B. VI]; (Anhydrous Stibious or Antimonious Acid—Sb_2O_3).......	lb. 1.50
N. B.—*The above* is the *Wet*-process Tri-oxide; the *Dry*-process Tri-oxide is the so-called "Flowers of Antimony."	
" " do., with Calcium Phosphate,—(James's Febrile Powder),—see Antimonial Powder, *U. S. Ph.* .	
" " brown,—*so-called*,—*washed*, (*Crocus* [Saffron] of Antimony; Crocus metallorum), see Potassa, antimonio-sulphurated, *washed*........	
" " " —*so-called*, — *unwashed*, (*Liver* of Antimony), see Potassa, antimonio-sulphurated, *crude*.........	
" oxy-chloride, (Powder of Algaroth)....	oz. .35
" oxy-sulphuret, Antimonious, (Kermes Mineral), see Antimony, sulphide, red,—so-called................	
" per-chloride (penta-chloride), [Antimonic chloride]..................	oz. .40
" sulphate	lb. 1.25
" sulphide, golden, (Antimonic sulphide, Penta-sulphide of Antimony), [so-called "Golden Sulphur"], I, chem. pure........	lb. 1.00
" " " II ....:	lb. .90
" " " III	lb. .75
" " black, (Antimonious sulphide, Tri-sulphide of Antimony), [Black Antimony], levigated, I,—pure;—*Antimonii sulphidum purificatum, U. S. Ph.*	lb. .50
" " " levigated,II,—*Antimonii sulphidum, U. S. Ph.*.......	lb. .35
" " " chem. pure, —synthetically prepared,—Ph. Gall....	lb. 2.00
" " vitreous,—so-called,—(Antimonial Glass; Vitreous Antimony)....	lb. .75
" " red,—so-called,—(Antimonious Oxy-sulphuret), [Kermes Mineral], (Red Antimony).	lb. 1.25
" " " Ph. G. I...................	lb. 1.75
" " " according to Cluzel........	lb. 2.00
" tannate..............................	

☞ When ordering, specify : "**MERCK'S**" !

	Containers incl.		
Antimony, tartarated (*tartarized*), [*Tartar Emetic*], (Antimoniated [Stibinated] Tartar), see Antimony and Potassium, tartrate, *U. S. Ph.*; and other grades			
" tartrate.—(*Do not confound* with above!)	oz. .35		
Antimony, black, see Antimony, *sulphide, black*			
" **red,** see do., *do.*, red,—so-called			
" **vitreous,** see do., *do.*, vitreous, so-called			
Antimony and **Mercury** Sulphides (Black Sulphides [Sulphurets]), see Mercury, antimonio-sulphide			
" and **Potassium,** oxalate, cryst.	lb. .75		
" " " tartrate, (Tartar Emetic; Tartarated [Tartarized] Antimony*), [Tartarus stibiatus—Antimoniated Tartar], cryst.	lb. .65		
" " " do., powder.	lb. .65		
N.B.—*Both the above preparations are of full percentage,*-abt. 43% Sb_2O_3.			
" " " do., pure, cryst.	lb. 1.00		
" " " " " powder,—*U. S. Ph.*, Ph.G.II,& Ph.Au.	lb. 1.00		
* N.B.—TARTARATED ANTIMONY *should not be confounded with:* Antimony, tartrate!			
Antimony, Butter of, *liquid*, see Solutions: Antimonious chloride			
" do. do., *concentrated,* see Antimony, chloride, Antimonious, etc.			
" **Crocus** (Saffron) of, [so-called "*Washed* Brown Oxide of Antimony"], see Potassa, antimonio-sulphurated, *washed*			
" **Flowers** of,—see remark under Flowers of Antimony.			
" **Glass** of, see Antimony, sulphide, vitreous,—so-called			
" **Liver** of, (so-called "*Unwashed* Brown Oxide of Antimony"), see Potassa, antimonio-sulphurated, *crude*			
" do. do., **calcic,** (also called: *Antimoniated* Liver of *Lime*), see Lime, antimonio-sulphurated			
Anti-Phylloxerins, see Potassium: sulphocarbonate; and, xanthogenate			
Antipyrine (Di - methyl - oxy - quinizine [-chinizine]);—also called "*Analgesine*".	oz. 1.40		
Apiol, *fluid,* green.. } Oily substances fom the seeds of	oz. .65		
" " distilled } Parsley (Apium petroselinum)..	oz. 1.50		
Apiol, solid, cryst., white, (Parsley-camphor)	15 gr. .25		
Apo-codeine	15 gr. 2.50		
" **hydrochlorate**	15 gr. 2.50		
Apocynin, cryst. } *Resinoid,—not* identical with	15 gr. 5.00		
" amorphous } the Glucoside "*Apocynin*"!	15 gr. 3.00		
Apo - morphine (Apomorphia), **hydrochlorate, amorphous**	⅛ oz.vls.oz. 5.25		
• " **hydrochlorate, cryst., chem. pure,—** *U. S. Ph.*	⅛ oz.vls.oz.11.75		
" **sulphate, cryst.,**—soluble in Water	15 gr. 2.00		
Apple-oil, so-called, see Amyl, valerianate.			
Aqua (Aquæ medicatæ—Medicated Waters), see Water, etc.			
Aqua Ammoniæ, see Ammonia, Water of.			
" **Calcariæ,** see Solutions: Lime, *U.S.Ph.*			
" **carmelitana,** see Spirit, Balm,—compound			
" **regia,** see Acid, nitro - hydrochloric, *U. S. Ph.*			

☞ When ordering, specify: "**MERCK'S**"!

	Containers incl.		
Arabin (Arabic Acid, Gummic Acid)........	oz. 1.00		
Arbutin Merck, white, cryst.	oz. 1.75		
Argentum, and compounds, see Silver, etc.			
Argil (Argilla) [Alumina], anhydrous, chem. pure, see Aluminium, oxide, anhydrous...............................			
" hydrated,—commercial; and: pure, U.S. Ph.;—see Aluminium, oxide, precipitated, etc.; etc.			
Arnicin	15 gr. 2.00		
Arsenic (Arsenium),—so-called "metallic,"—cryst.;[so-called "Cobaltum Mineral"]	oz. .12		
" bromide.............................	oz. .50		
" chloride.............................	oz. .60		
" iodide (ter-iodide), cryst., pure,—U.S.Ph.	oz. .60		
" " with Mercury bin-iodide, see Mercury, arsenio-iodide			
" lactate, (Lacto-arsenious Acid)..........	oz. 2.50		
" oleate................................	oz. .40		
" pent-oxide, see Acid, arsénic (arsenicic), dry [anhydrous]..............			
" " tetra-hydrated, see Acid, arsénic (arsenicic), hydrated..........			
" phosphide (phosphuret)..............	oz. 1.00		
" Red sulphide, (di-sulphide), [Realgar; Red Arsenic], powder.............	lb. .25		
" tartrate	oz. .40		
" tri-oxide, see Acid, arsenious (arsenicous), anhydrous			
" Yellow sulphide, (tri-sulphide), [Yellow Arsenic, Citrine Arsenic; Orpiment — Auri Pigmentum; King's Yellow], powder..........	lb. .25		
" " " precipitated (wet process).	oz. .35		
Arsenic, red, see Arsenic, Red sulphide....			
" vitreous, } see Acid, arsenious, lumps " Glass of,			
" white, so-called....... } see Acid, arsenious, etc. " Flowers of, resublimed			
" yellow (citrine), see Arsenic, Yellow sulphide.............................			
Arsenic and Mercury Iodides, see Mercury, arsenio-iodide........................			
do. do. do. do., solution, U. S. Ph., (Donovan's Solution), see under Solutions			
Arsenical Solution, Fowler's, see Solutions: Potassium arsenite, U. S. Ph............			
Arsenium, and compounds, see Arsenic, etc.			
Asaron (Asarin; Asarum-camphor; Asarabacca-camphor)	15 gr. .75		
Aseptol (ortho-Phenol-sulphonic [ortho-Phenyl-sulphuric, ortho-Sulpho-phenic, ortho-Sulpho-carbolic] Acid; ortho-Sulphophenol [-carbol]; — in 33⅓-% solution) — [Sozolic Acid]	oz. .30		
Asparagin (Asparagine; Althein, Altheine).	oz. 1.00		
Aspidos-amine and Aspido-spermine, see under Quebracho Alkaloids			
Atropine Merck (Atropia):			
pure, heavy,—Atropina, U. S. Ph.—Alkaloid from Atropa Belladonna,—free from the so-called "light Daturine."—Melt.-point 115° C [239 F]	⅛ oz.vls.oz. 6 .55		
arseniate (arsenate)	15 gr. .65		
borate	15 gr. .50		
hydrobromate	15 gr. .65		
hydrochlorate	15 gr. .65		
nitrate	15 gr. .65		
salicylate.............................	15 gr. .65		

☞ When ordering, specify: "MERCK'S"!

	Containers incl.
Atropine Merck (Atropia),—*continued*:	
santoninate (*not* santonate!)	15 gr. .75
sulphate, white, cryst., neutral, — *Atropinæ sulphas, U. S. Ph.*,—absolutely neutral (free from *any trace* of either acid or alkaline reaction!),-light, and perfectly white	⅛ oz. vls.oz. 5.70
tartrate	15 gr. .65
valerianate	15 gr. .65
N.B.—Atropine *fractional derivatives*, including **Hom-atropine Merck-Ladenburg**, will be found under their respective names.	
Atropine Discs,—in tubes of 100.........	
" **Gelatin**,—in sheets for 25 applications.	
" **Paper**,—in books for 100 applications.	
Auri Pigmentum, see Arsenic, Yellow sulphide.................................	
Aurin, see under Aniline and Phenol Dyes..	
Auro- double salts, see "Gold and—"......	
Aurum, and compounds, see Gold, etc.....	
Avenin - Legumin (*Vegetable Casein* from oats).....................................	oz. 1.00
Avenine,—*Alkaloid*	15 gr. .60
Azo-benzene (Azo-benzol, Azo-benzide)....	oz. 1.25
Azo-litmin, chem. pure....................	15 gr. .75

When ordering, specify: "**MERCK'S**"!

Containers incl.

☞ When ordering, specify: "MERCK'S"!

	Containers incl.
Balsams:	
Copaiva, Maracaibo-, —(Balsamum capivi [copaivæ])	lb. 1.00
" —dry, — (Balsamum copaivæ siccum), see Resins: Copaiva	
Gurjun, —(so-called "*East-India Copaiva Balsam*"), [also called: "Wood-oil," or "East-Indian Wood-oil"]	lb. ⅔ .75
Indian Hemp,—(Balsamum cannabis indicæ),—acc. to Denzel	oz. 2.50
Kava-Kava, see **Resins: Kava-Kava**	
of Peru, true	lb. 2.50
of Sulphur, see Oils, divers: sulphurated Linseed-	
" " terebinthinated, see Oils, divers: sulphurated Linseed-, terebinthinated	
Bamberger's Solution, mercuro-albuminated, see Mercury, bi-chloride, albuminated, fluid	
Baptisin, pure.—Glucoside from Wild Indigo, (Baptisia tinctoria)	15 gr. .50
Barb-aloin, chem. pure, see **Aloin**	
Barium (Baryum, Barytum), double salts of, see "Barium and —" (below!)	
" metallic	15 gr. 4.00
" acetate, pure, cryst.	oz. .20
" " chem. pure, cryst.	oz. .25
" æthylo-sulphate, see Barium, ethylo-sulphate	
" amylo-sulphate	oz. .40
" anhydride, so-called, see Barium, oxide, anhydrous, pure; and, commercial	
" benzoate	oz. .75
" bi-oxalate (bin-oxalate)	lb. 1.25
" borate	oz. .40
" boro-wolframate (boro-tungstate)	
" bromate	oz. .60
" bromide	oz. .35
" carbonate, precipitated	lb. .40
" " " pure	lb. .55
" " " chem. pure	lb. 1.00
" chlorate, pure, cryst.	lb. .70
" " " powder	lb. .75
" chloride, impalp. powder, commercial	lb. .25
" " purified, cryst.	lb. .25
" " pure, cryst.	lb. .30
" " chem. pure, cryst.	lb. .35
" chromate, pure	lb. 1.00
" " II	lb. .60
' citrate	oz. .40
" ethylo-sulphate (sulpho-vinate), cryst.	lb. 2.25
" fluoride, pure	oz. .75
" formate	oz. .75
" hydroxide (so-called "hydrate") [hydrated mon-oxide], see Barium, oxide, hydrated, etc.	
" hypo-phosphite	oz. .50
" hypo-sulphate	oz. .60
" hypo-sulphite, see Barium, thio-sulphate	
" iodate	oz. 1.00
" iodide	oz. .75
" lactate	oz. .85
" methylo-sulphate	oz. .60
" nitrate, cryst.	lb. .25
" " powder	lb. .25
" " fused	lb. 1.00
" " chem. pure, cryst.	lb. .45

☞ **When ordering, specify: "MERCK'S"!**

	Containers incl.
Barium, oleate	oz. .40
" oxalate	lb. .75
" " pure	lb. 1.00
" oxide (mon-oxide), anhydrous, [Burnt (calcined) Baryta], (so-called "Barium Anhydride"), pure	lb. 2.50
" " do., commercial	lb. 1.50
" " hydrated (caustic), [Barium Hydroxide (so-called "Hydrate"); Hydrated (caustic) Baryta], pure, cryst.	lb. .50
" " do., pure, dry	lb. 1.00
" " " chem. pure, cryst.	lb. .60
" " " " " dry	lb. 1.25
" " " commercial	lb. .40
" oxide, per- (di-), see Barium, per-oxide.	
" per-chlorate	oz. 2.00
" per-manganate, cryst.	oz. 1.50
" per-oxide (di-oxide), hydrated, pure	lb. 1.25
" " do., commercial	lb. .75
" " anhydrous, commercial	lb. .75
" " " pure	lb. .85
" phosphate	oz. .35
" rhodanide, see Barium, sulpho-cyanate.	
" salicylate	oz. .50
" sulphate, precipitated, pure,–(Synthetically prepared "Barytes"; also called: Artificial "Heavy Spar")	lb. .55
" sulphide (sulphuret), commercial	lb. .45
" " pure	lb. 1.00
" " — free from Arsenic; — acc. to Winkler.—(Used for generating Arsenium-free Sulphydric Acid in Kipp's apparatus.)	lb. .60
" sulpho-carbolate (phenol-sulphonate, sulpho-phenate)	lb. 1.75
" sulpho-cyanate (thio-cyanate; rhodanide), pure	lb. 1.50
" " commercial	lb. .75
" sulpho-vinate, see Barium, ethylo-sulphate	
" tartrate	oz. .75
" thio-cyanate, see Barium, sulpho-cyanate	
" thio-sulphate (formerly called "hypo-sulphite")	oz. .30
" wolframate (tungstate)	oz. .30
Barium and Platinum, salts, see Platinum double Chlorides; do. double Cyanides; and do., divers double Salts	
" **and Potassium,** chlorate	lb. 1.50
Baryta, burnt (calcined), see Barium, oxide, anhydrous	
" caustic (hydrated), see Barium, oxide, hydrated	
Barytes, synthetically prepared, (Artificial "Heavy Spar"), see Barium, sulphate, etc.	
Bebeerine, (Beberine, Beeberine, Bibirine, Bebeeria; Buxine), pure, cryst.	oz. 1.65
" hydrochlorate	oz. 1.25
" sulphate	oz. 1.25
Belladonnine	15 gr. .75
Bengal Rose, see under Aniline and Phenol Dyes: Rose	
Benz-aldehyd (Benzoic Aldehyd; so-called "Benzoyl Hydride") [*Artificial* Volatile Oil of Bitter Almonds;—*not*=Nitro-benzene!—*which see also!*].—Chemically identical with: De-hydrocyanated *Natural* Essential Oil of Bitter Almonds	lb. 2.00

☞ **When ordering, specify: "MERCK'S"!**

	Containers incl.
Benz-amide............................	oz. 2.00
Benzene (Benzol), bromated, see Mono-brom-benzene...................	
" chlorated, see Mono-chlor-benzene....	
" iodated, see Mono-iod-benzene........	
Benzene, anthracic, (Coal-tar Benzol), [Coal-naphtha; so-called "Coal-tar Benzin"—Benzinum lith-anthracinum],—chem. pure, crystallizable;—boil.-pt. 80-84° C [176-183.2 F].— (So-called "Phenyl Hydride.")....'.	lb. 1.00
" do.,—boil.-pt. 70-130° C [158-266 F]..	lb. .75
" do.,— " 130-180° C [266-356 F]..	lb. .50
Benzene, benzoic, see Benzol, benzoic....	
Benzid-am, see Aniline..................	
Benzile (Di-benzoyl).....................	15 gr. .75
Benzin, petroleic, (Petroleum Benzin), [Petroleum Naphtha],—I,— boil.-pt. 55-75° C [131-167 F]................	
" do.,—boil.-pt. 50-60° C [122-140 F],—Benzinum, U. S. Ph.,—(so-called "Petroleum Ether")..................	
Benzo- (Benzene-) [Benzol-] Quinone, see Quinone............................	
Benzo-tri-chloride (not Tri-chlor-benzene; nor Tri-chloride of Benzene [Benzol];-but: $C_6H_5.CCl_3$)...................	oz. .50
Benzoin Crystals, (Bitter-almond-oil Camphor), [not: Resina Benzoë, = "Gum benjamin"; — but: Oxy-phenyl-benzyl-ketone!]..	15 gr. .35
Benzoin Flowers, see Acid, benzoic, from Siamese (and other) Benzoin-resin, sublimed: U. S. Ph., and others.....................	
Benzol (Benzene), bromated, see Mono-brom-benzene.....................	
" chlorated, see Mono-chlor-benzene....	
" iodated, see Mono-iod-benzene......	
Benzol, benzoic, (Benzoic Benzene),—from Benzoic Acid.........................	oz. 1.50
Benzol of Coal-tar, (Anthracic Benzol), see Benzene, anthracic......................	
Benzoyl, chloride.......................	oz. .50
" hydride,—so-called,—see Benz-aldehyd	
Benzoyl, di-, see Benzile................	
Benzoyl-ecgonine.......................	15 gr. 1.50
Benzyl, chloride, commercial	lb. 1.50
" " pure	lb. 3.00
Berberine, chem. pure, cryst.................	oz. 5.00
" citrate............:...........	15 gr. .75
" hydrochlorate	oz. 2.00
" phosphate.......................	15 gr. .75
" sulphate.......................	oz. 1.25
Berberine, Hydro-......................	15 gr. 4.00
Beryllium (Glucinum, Glycium), metallic, powder	15 gr.12.00
" carbonate	15 gr. .25
' chloride.........................	15 gr. .25
" oxide, hydrated, (hydroxide)..........	15 gr. .25
" " anhydrous	15 gr. .50
" sulphate	15 gr. .25
Beryllium and Potassium, fluoride.......	15 gr. .25
Bestuscheff's Solution, tonico-nervine (anodyne Iron-), see Tinctures: Iron chloride,—ethereal	
Betol (Naphthalol) [Naphtho-salol, Sali-naphthol]—(Beta-Naphthylic Ether of Salicylic Acid; Salicylate of Beta-Naphthol)........	oz. .60
Bibirine, see Bebeerine.................	
Bi-chlor-naphthalene, see Di-chlor-naphthalene............................	

☞ When ordering, specify: "MERCK'S"!

		Containers incl.			
Bili-fuscin.............................		1½ gr.vial 4.00			
Bili-humin.............................		1 gr.vial 2.00			
Bili-prasin.............................		1½ gr.vial 4.00			
Bili-rubin (Bili-phain).................		1½ gr.vial 4.00			
" Hydro-, see Uro-bilin..............		.....			
Bili-verdin............................		1½ gr.vial 4.00			
Bi-methyl- compounds, see Di-methyl- etc.					
Bi-nitro-benzene, (Bi-nitro-benzol, Bi-nitro-benzide), see Di-nitro-benzene.......					
Bi-nitro-naphthalene, see Di-nitro-naphthalene.................................					
Bi-nitro-toluene (-toluol), see Di-nitro-toluene................................					
Bi-phenyl- and other Bi-compounds, see Di-phenyl- etc.;—etc.,—under "Di-"—					
Birch-tar, see Oils, divers: Birch, empyreumatic..............................					
Bismarck Brown, see under Aniline and Phenol Dyes: Brown					
Bismuth, double salts of, see "Bismuth and —" (below!)....................					
" metallic,—about 97% pure metal.....		lb. 2.40			
" " pure,—free from Arsenic........		lb. 3.50			
" " " chem. pure		lb. 6.00			
" acetate............................		oz. .60			
" albuminate........................		oz. .60			
" ammonio-citrate, see Bismuth and Ammonium, citrate, U. S. Ph.					
" benzoate		oz. .60			
" bromide		oz. 1.00			
" camphorate		oz. 2.00			
" carbonate, so-called, see Bismuth, sub-carbonate, U. S. Ph.					
" chromate		oz. .75			
" citrate,—U. S. Ph...................		oz. .50			
" iodide (ter-iodide)...................		oz. .80			
" lactate		oz. 1.00			
" lacto-phosphate (phospho-lactate).....		oz. 1.00			
" nitrate, cryst.		lb. 2.50			
" oleate, dry........................		oz. .35			
" oxalate............................		oz. .50			
" oxide (tri-[sesqui-]oxide), anhydrous [yellow], chem. pure,—Ph.Brit. .		oz. .60			
" " hydrated (white), pure		oz. .50			
" oxide, per- (pent-), see Bism., per-oxide					
" oxy-chloride		oz. .35			
" oxy-iodide (sub-iodide).............		oz. .55			
" peptonized, (Bismuthated Peptone),—contains 3.8% of Oxide of Bismuth in soluble form......................		oz. .75			
" per-manganate, basic,—soluble only in dilute acids........................		oz. 1.75			
" per-oxide (pent-oxide)................		oz. .75			
" phosphate.........................		oz. .60			
" phospho-lactate, see Bismuth, lacto-phosphate..........................					
" salicylate, basic,—contains about 62% of Bi_2O_3,—free from the Sub-nitrate;—gives up only traces of Salicylic Acid to Ether		oz. .45			
" salicylate, acid,—contains about 40% of Bi_2O_3,—free from the Sub-nitrate...		oz. .40			
" sub-carbonate,—U. S. Ph.,—(so-called "carbonate"),—chem. pure..........		lb. 2.90			
" sub-iodide, see Bismuth, oxy-iodide.......					
" sub-nitrate, chem. pure, very light powder, — U. S. Ph. and Ph. G. II,—(Magistery of Bismuth);— perfectly free from Arsenic, by Marsh's test............		lb. 2.50			
" sub-nitrate, in tablets		lb. 2.75			

☞ When ordering, specify: "MERCK'S"!

	Containers incl.
Bismuth, sulphate	oz. .50
" sulphide (sulphuret)	oz. .60
" tannate	oz. .35
" " in tablets	oz. .40
" tartrate	oz. .75
" valerianate	oz. .75
Bismuth and **Ammonium**, citrate, — U. S. Ph.	oz. .50
" and **Potassium**, iodide, liquid	oz. .60
" " " tartrate "	oz. .25
Bitter-almond-oil, artificial, see Benzaldehyd	
Bitter-almond-oil Camphor, see Benzoin Crystals	
Bixin (*Red* Orellin), chem. pure	oz. 5.00
Blood, bullock's, (Sanguis Tauri [Bovis]), dry, powdered	lb. 1.50
Boldine	15 gr. 3.00
Bone-ash; and: do., purified;—see Calcium, phosphate, crude; and: pure. N.B.—*Compare, also:* Calcium, phosphate, bi-basic,—for agricultural chemistry.	
Bone-black, purified, (so-called "Ivory-black"), see Charcoal, animal, purified, *U. S. Ph.*,—etc.	
Bone Phosphate,—so-called,—see Calcium, phosphate, precipit'd tri-basic, dry, *U. S. Ph.*	
Borax,—*various forms*, (also: Borax-glass), —see Sodium, bi-borate, etc.,—*U. S. Ph.*; and other forms	
Borax-Tartar (*so-called* "Soluble Cream of Tartar"), see Potassium and Sodium, boro-tartrate	
do. Scales, (*Scales of Tartar*),—*perfectly soluble* in Water,—see do. do. do., do.,—*in scales*	
Boro-Glycerin ("-Glyceride"), *dry*,—[Glycerolate (Glycerite) of Boric Acid; Glyceryl Borate];—containing 3 parts Glycerin to 2 of Boric Acid	lb. 2.00
" so-called, — *syrupy consistency;* — see Sodium, bi-borate, glycerolate of,—*syrupy consistency*	
Boron (Borium), crystallized	15 gr. 6.00
Brayerin, see **Koussein Merck**	
Bromal, anhydrous	oz. 2.50
Bromal Hydrate	oz. 2.50
Bromine,—*Bromum, U. S. Ph.*	oz. .25
" chloride, ("Bromide of Chlorine," so-called)	oz. .85
" iodide, liquid,—so-called,—see Iodine, bromide, liquid	
Bromo-Caffeine (*not* Caffeine Hydrobromate, — *which see also;* — but Bromated [bromo-substituted] Caffeine!)	oz. 5.00
Bromo-ethyl (Bromide of Ethyl; Mono-bromethane), see **Ether, hydrobromic**	
Bromoform	oz. 1.50
Brom-phenyl-acet-amide, mono-, (Monobrom-acet-anilide), cryst. — [Supposed to combine the medicinal effects of Sodium Bromide and of Phenyl-acet-amide.]	oz. 2.00
Brucine (Brucia) [Vomicine], chem. pure, cryst.,—free from Strychnine	¼ oz. vls. oz. 3.00
" pure	⅛ oz. vls. oz. 2.10
" hydrobromate	⅛ oz. vls. oz. 2.10
" hydrochlorate	⅛ oz. vls. oz. 2.10
" nitrate	⅛ oz. vls. oz. 2.10
" phosphate	¼ oz. vls. oz. 3.50
" sulphate	⅛ oz. vls. oz. 2.10

☞ **When ordering, specify: "MERCK'S"!**

	Containers incl.
Brucine and Zinc-Oxide, hydriodate	15 gr. 1.50
Bryonin................................	15 gr. .50
Butter, Cacao-, (Oil of Theobroma), *fresh*.	lb. .75
" Nutmeg-, see Oils, divers: Nutmeg, expressed	
Butter of Antimony, *liquid*, see Solutions: Antimonious chloride	
" of do., *concentrated*, see Antimony, chloride, Antimonious.................	
" of Tin, anhydrous, see Tin, tetrachloride	
" of Zinc, see Zinc, chloride: *U. S. Ph.* forms; and others	
Butyl Iodide...........................	oz. 3.00
Butyl-chloral Hydrate (Croton-chloral Hydrate).	oz. .60
Butyl-phenol...........................	oz. 2.50
Butyrin (Tri-butyrin).....................	15 gr. .35
Butyrum stibii (antimonii); do. myristicæ (nucistæ); do. stanni; do. zinci;—see references under: Butter of Antimony; of Nutmeg; of Tin; of Zinc.—*Butyrum Cacao*, see Butter, Cacao-	
Buxine, see Bebeerine.....................	

☞ When ordering, specify: "MERCK'S"!

	Containers incl.			
Cacao-butter, see Butter, Cacao-				
Cadmium, double salts of, see "Cadmium and —" (below!)				
" metallic	lb. 1.45			
" " sheet	lb. 3.00			
" " powder	oz. .75			
" acetate	oz. .75			
" boro-wolframate (boro-tungstate), solution,—sp. gr. 3.28	oz. 1.00			
" bromide	oz. .27			
" carbonate	oz. .50			
" chlorate	oz. .75			
" chloride	oz. .35			
" fluoride	oz. 1.00			
" iodide	oz. .45			
" nitrate	oz. .40			
" oxide	oz. .60			
" salicylate	oz. 1.50			
" sulphate, pure	oz. .30			
" sulphide (sulphuret)	oz. .50			
" sulpho-carbolate (phenol-sulphonate, sulpho-phenate)	oz. .75			
" tartrate	oz. .75			
" valerianate	oz. 1.00			
Cadmium and Ammonium, bromide	oz. .50			
" " " iodide	oz. .60			
" and **Gold,** chloride, see Gold & C., chlor.				
" and **Potassium,** iodide	oz. .60			
Caesium (Cæsium), metallic				
" bi-tartrate	15 gr. 3.00			
" chloride	15 gr. 3.50			
Caesium and Rubidium, chloride	15 gr. 2.00			
Caesium Alum, see Alum, caesic				
Caffeine (Caffeia, Coffeine) [Theine], double salts of, see "Caffeine and —" (below!)				
" **pure, cryst.,— U. S. Ph.**	⅛ oz. vls. oz. 1.00			
" **pure, true,—from Coffee-seeds**	⅛ oz. vls. oz. 4.00			
" **acetate, true salt**	⅛ oz. vls. oz. 3.00			
" **ammonio-citrate,** see **Caffeine** and **Ammonia, citrate**				
" **arseniate** (arsenate)	⅛ oz. vls. oz. 3.00			
" **arsenite**	⅛ oz. vls. oz. 3.00			
" **benzoate, true salt**	⅛ oz. vls. oz. 1.75			
" **boro-citrate,** true double salt,—readily soluble.—(Combines the medicinal effects of Caffeine and of Boric Acid.)	⅛ oz. vls. oz. 3.50			
" **bromo-substituted (bromated),** [not Caffeine Hydrobromate!], see Bromo-Caffeine				
" **carbolate,** see Caffeine, phenate				
" **cinnamate, cryst**	⅛ oz. vls. oz. 4.00			
" **citrate, true salt**	⅛ oz. vls. oz. 1.75			
" **citrate,—so-called,—commercial**	⅛ oz. vls. oz. 1.00			
" " " —Ph. Brit. new,—[50% of Caffeine]	⅛ oz. vls. oz. 1.00			
" **citrico-benzoate**	⅛ oz. vls. oz. 2.00			
" **hydrobromate, true salt, cryst.** N. B.—Compare, also: Bromo-Caffeine.	⅛ oz. vls. oz. 1.20			
" **hydrochlorate, true salt, cryst.**	⅛ oz. vls. oz. 1.20			
" **lactate**	⅛ oz. vls. oz. 1.75			
" **malate**	⅛ oz. vls. oz. 5.00			
" **nitrate, true salt, cryst.**	⅛ oz. vls. oz. 2.00			
" **phenate** (phenylate, carbolate)	⅛ oz. vls. oz. 3.50			
" **phtalate,**—soluble in 5 parts of Water.	⅛ oz. vls. oz. 3.00			
" **salicylate, true salt**	⅛ oz. vls. oz. 1.90			
" **sodio-hydrobromate,**—and other Soda double salts of Caffeine,—see **Caffeine and Soda,** etc. -(below!)				
" **sulphate, true salt, cryst.**	⅛ oz. vls. oz. 1.90			

☞ When ordering, specify: "**MERCK'S**"!

		Containers incl.
Caffeine—(*as above!*),—tannate, true salt ...		⅛ oz.vls.oz. 2.00
" valerianate, true salt		⅛ oz.vls.oz. 2.00
Caffeine and Ammonia, citrate, (Ammoniated Citrate of Caffeine) .. [54% of Caffeine]		¼ oz.vls.oz. 2.00
" and Soda, benzoate .... [45.8% "]		⅛ oz.vls.oz. 1.25
" " " cinnamate ... [62.5% " .]		⅛ oz.vls.oz. 1.50
" " " citrate, true .. [52.5% "]		¼ oz.vls.oz. 2.00
" " " salicylate. . . [62.5% "]		⅛ oz.vls.oz. 1.25

N.B.—The *Benzoate*, *Cinnamate*, and *Salicylate*, are soluble in 2 parts of hot Water, and remain in solution on cooling.

Caffeine and Soda, hydrobromate, (so-called "Bromide of Caffeine and Sodium",— "Sodio-bromide of Caffeine"),—[52% of Caffeine];—soluble in 20 parts of Water .. ⅛ oz.vls.oz. 1.50

Cahincin (Caincin), see Acid, cahincic......
Calabar-Alkaloid, see Physostigmine (Eserine) ...
Calabar Discs........ ⎫
" Gelatine .... ⎬ see Physostigmine Discs; etc.; etc.
" Paper.......... ⎭

Calcium, double and triple salts of, see "Calcium and —" (below!)..............
" metallic,—by electrolysis............. 15 gr. 10.00
" acetate, chem. pure, dry............ lb. 1.00
" " crude................. lb. .50
" æthylo-sulphate, see Calc., ethylo-sulph.
" albuminate.......................... oz. .75
" antimonio-sulphide, so-called, (Antimonic Liver of Lime), see Lime, antimonio-sulphurated..................
" arseniate (arsenate)................... oz. .35
" arsenite .... oz. .30
" benzoate .... oz. .50
" bi-malate, cryst..................... oz. 1.00
" bi-saccharate, see Calcium, saccharate..
" bi-phosphate, so-called, see Calcium, phosphate, acid....................
" bi-sulphate, pure.................... oz. .30
" bi-sulphite, liquid,—[8° Baumé]...... lb. .45
" bi-tartrate, pure .... oz. .40
" borate .... lb. 1.50
" " —glycerolate (glycerite) of, [Glycerino-borate of Calcium)....... lb. 2.50
" boro-citrate........................ oz. .35
" bromide,—*U. S. Ph.*................ oz. .25
" bromo-iodide....................... oz. 1.00
" butyrate, pure..................... oz. .50
" " Iso-, see Calcium, iso-butyrate ...
" carbolate, see Calcium, phenate.........
" carbonate, purified (elutriated), white, see Chalk, prepared, *U. S. Ph.*
" " precipitated................... lb. .40
" " " light (flocculent) .... lb. .45
" " " pure,—*U. S. Ph.* & Ph. G. II. lb. .50
" " " chem. pure.............. lb. 1.00
" chinate, see Calcium, quinate.........
" chinovate, see Calcium, quinovate.....
" chlorate............................ oz. .40
" chlorhydro-phosphate, see Calcium, phosphate, hydrochlorated...........
" chloride, (so-called "Hydrochlorate of Lime"), crude......................
" " granulated...................... lb. .35
" " pure, cryst.................... lb. .40
" " " dry, white................ lb. .45
" " " fused, perf. white, in lumps, —*U.S.Ph.* lb. .65
" " " " " " in sticks.. lb. .85
" " " " " " granulated lb. .90
" chromate......................... lb. 2.50

☞ When ordering, specify: "MERCK'S"!

	Containers incl.		
Calcium, citrate	oz. .35		
" ethylo-sulphate (sulpho-vinate)	oz. .75		
" ferrid-cyanide (ferri-cyanide), [Calcio-Ferric cyanide, so-called], cryst.	oz. .50		
ferro-lacto-phosphate. (Lacto-Phosphate of Calcium and Iron)	oz. .50		
" formate	oz. .35		
" fluoride, chem. pure	lb. 2.50		
" glycerino-borate, (Glycerolate [Glycerite] of Borate of Calcium), see Calcium, borate,—glycerolate of			
" glycerino-phosphate, (Glycerolate [Glycerite] of Phosphate of Calcium), see Calcium, phosphate,—glycerolate of.			
" **hippurate**	oz. 2.00		
" hydrochloro-phosphate, see Calcium, phosphate, hydrochlorated			
" hypo-phosphite,—*U. S. Ph.*	lb. 1.30		
" hypo-sulphite, see Calc., thio-sulphate.			
" iso-butyrate	oz. 2.00		
" iodate	oz. .75		
" iodide	oz. .47		
" kinate, see Calcium, quinate			
" kinovate, see Calcium, quinovate			
" **lactate, pure, soluble**	oz. .25		
" **lacto-phosphate** (phospho-lactate), **cryst., soluble**	oz. .50		
" " powder	oz. .25		
" meconate			
" muriato-phosphate, see Calcium, phosphate, hydrochlorated			
" nitrate, pure	oz. .15		
" nitrite	oz. .25		
" oleate	oz. .45		
" **osmate**	15 gr. 2.50		
" oxalate	oz. .30		
" oxide, caustic, dry, (Burnt Lime, pure), —from marble,—see Lime, *U. S. Ph.*			
" per-manganate, cryst.	oz. 2.00		
" phenate (phenylate, carbolate), pure	lb. 1.50		
" " crude, [about 40% of pure]	lb. .40		
" phosphate, crude, (*Bone-ash*)	lb. .40		
" " pure, (*Purified Bone-ash*)	lb. .60		
" " neutral, chem. pure,—Ph. G. II,—(Tetra-hydrated Di-calcic ortho-Phosphate; Di-hydrated Calcium Hydro-phosphate)	lb. 1.25		
" " acid, (so-called "bi-phosphate"), [Tetra-hydro-mono-calcic ortho-Phosphate], pure	lb. 2.00		
" " bi-basic,—for agricultural chemistry	lb. 1.50		
" " precipitated tri-basic, dry,—*Calcii phosphas præcipitatus, U. S. Ph.,*—(so-called "*Bone Phosphate*")	lb. 1.25		
" " do. do., *gelatinous*	lb. .75		
" " —glycerolate (glycerite) of, [Glycerino-phosphate of Calcium]	oz. 4.00		
" " —hydro-chlorated (muriated), [Muriato-phosphate (Chlorhydro-phosphate, Hydrochloro-phosphate) of Calcium], liquid, —sp. gr. 1,225, [25% solution]	lb. .75		
" " —do., dry	lb. 1.50		
" " —antimoniated (stibiated),—[James's Febrile Powder],—see Antimonial Powder, *U. S. Ph.*			
" phosphide (phosphuret)	oz. .50		
" phosphite	oz. .75		
" **phospho-lactate,** see **Calcium, lacto-phosphate**			

☞ When ordering, specify: "**MERCK'S**"!

	Containers incl.			
Calcium, picrate (picro-nitrate)............	oz. .30			
" pyro-phosphate.....................	oz. .30			
" quinate (chinate, kinate), cryst.	oz. 1.00			
" quinovate (chinovate, kinovate).......	oz. 1.00			
" rhodanide, see Calcium, sulpho-cyanate				
" saccharate (bi-saccharate), [so-called "Saccharate of Lime"],—soluble in Water, easily so in Sugared water.— (Antidote in Carbolic-Acid poisoning.)	oz. .25			
" salicylate...........................	oz. .45			
" **santoninate** (*not* santonate!),—white powder, insoluble in water; insipid......	oz. .75			
" silicate, pure........................	oz. .35			
" silico-fluoride........................	oz. .40			
" stibiato-sulphide, so-called, (Antimonic Liver of Lime), see Lime, antimonio-sulphurated				
" sulphide (sulphuret),—acc. to Fresenius...... } For generating Sulphydric Acid in Kipp's apparatus.	lb. 1.00			
" " —acc. to Otto......	lb. 1.10			
" sulphide, *so-called*, (Calcic Liver of Sulphur), see Lime, sulphurated, *U. S. Ph.*				
" do., antimoniated,—*so-called*,—(Antimonic Liver of Lime), see Lime, antimonio-sulphurated..................				
" sulphite, crude	lb. .30			
" " purified.........................	lb. .50			
" " pure.............................	lb. 1.25			
" sulpho-carbolate (phenol-sulphonate, sulpho-phenate)..................	oz. .25			
" sulpho-cyanate (thio-cyanate; rhodanide), commercial...............	lb. .75			
" " pure	lb. 1.25			
" sulpho-vinate, see Calc., ethylo-sulphate				
" tannate	oz. .30			
" tartrate	oz. .25			
" thio-cyanate, see Calc., sulpho-cyanate				
" thio-sulphate (formerly called "hyposulphite").........................	lb. 1.25			
" tri-chlor-phenate (tri-chlor-phenylate, tri-chlor-carbolate)..................	oz. .50			
" urate, chem. pure	oz. 1.00			
Calcium and Copper, acetate, see Copper and Calcium, acetate...............				
" and **Gold,** chloride, see G.& Calc., chlor.				
" and **Iron,** lacto-phosphate, see Calcium, ferro-lacto-phosphate......				
" " " cyanide, so-called, see Calcium, ferrid-cyanide........				
" and **Platinum,** cyanide, see under Platinum double Cyanides				
Calcium, Platinum, and **Ammonium,** cyanuret, see under Platinum triple Cyanides				
Calomel, see Mercury, chloride, *U.S.Ph.;* etc.				
Calx, *U. S. Ph.,* see Lime, *U. S. Ph.*.......				
Calx Antimonii (*Stibii*), see Potassium, antimonate, *pharmacopeial*.......				
" " **cum Sulphure,** see Lime, antimonio-sulphurated				
Camphor, benzoated....................	oz. .40			
" carbolated, see Camphor, phenolated..				
" citrated..............................	oz. .40			
" di-bromated........................	oz. 1.00			
" mono-bromated,—*U. S. Ph.*...........	oz. .26			
" phenolated, (*Phenol-Camphor;* Carbolated camphor, Camphorated Phenol).	oz. .40			
" salicylated	oz. .50			
" valerianated.......................	oz. .60			
Camphor, artificial, so-called, see Turpentine-oil, mono-hydrochlorate				

☞ When ordering, specify : "**MERCK'S**" !

MERCK'S INDEX.

	Containers incl.
Camphor of **Anemone** (*Pulsatilla*), see Anemonin.............................	
" of **Asarum** (Asarabacca), see Asaron ..	
" of **Bitter-Almond-oil**, see Benzoin Crystals.............................	
" of **Elecampane** (*Inula, Alant-root*),–solid, —see **Helenin**............	
" " —*liquid*,—see Alantol......	
" **Lemon-**, so-called, see Turpentine-oil, di-hydrochlorate................	
" of **Parsley**, see Apiol, solid, cryst., white....	
" of **Thyme**, see Thymol	
" of **Tonka-bean**, see Cumarin........	
Cannábin,—Resinoid ....●...........	15 gr. .35
Cannabine Merck,—pure Alkaloid;—syrupy consistency.—(Simply hypnotic in action.)...	15 gr. 5.00
Cannabine Tannate Merck....................	15 gr. .25
Cannabinon.................................	15 gr. .20
" —10-% abstract in Sugar of milk,— adapted for immediate dispensation..	oz. .60
Cantharidin, cryst...........................	15 gr. 2.00
Capro-nitrile, see Amyl, cyanide..........	
Capsicin....................................	15 gr. .20
Caput mortuum, pure, see Iron, oxide, red, *anhydrous*	
Carb-amide, etc., see Urea, etc.............	
Carb-azole (Di-phenyl-imide)	oz. 1.00
Carbinol, see Alcohol, methylic...........	
Carbo animalis (Ossium), *purificatus,* U. S. Ph.;—et, purus............ ⎫	
" **Carnis** purus—ad usum *internum*. ⎪	
" **Sanguinis**; et,—acido purificatus. ⎪ see Charcoal, etc.	
" **Spongiæ** pulverisatus .... ⎬	
Carbon, animal (Bone-), purified, *U. S. Ph.*;—and, pure .... ⎪	
" **Blood-**; and: do., purified by acid. ⎪	
" **Meat-,** pure,—for *internal* use .... ⎪	
" **Sponge-,** powder.................. ⎭	
Carbon, mineral, see Graphite.............	
Carbon (*Carboneum*), bi-sulphide, (so-called Sulphur-"Alcohol")................	lb. .25
" do., highly rectified,—*U. S. Ph*........	lb. .40
" di-chloride (also called: proto-chloride).	
" tetra-chloride (also called: bi-chloride) .	oz. .75
" tri-chloride (also called: sesqui-chloride), cryst..........................	oz. 1.25
Cardol, pruriginous; from Anacardium orientale...........................	oz. .75
" vesicatory; from do. occidentale.......	oz. 1.00
Carica papaya, Juice of, see Juice of Papaw.....	
Carmine, pure, in lumps, (Nacarat).........	oz. .75
Carmine, Safflower-, see Safflower Carmine	
Carnine	15 gr. 8.00
" hydrochlorate	15 gr. 8.00
Carthamin, (so-called "Carthamic Acid"), chem. pure	15 gr. 1.00
Carvol, see Essential Oils: Caraway-seed; extra strong	
Casein, commercial ...... ⎫ *From milk* ⎧ ..	lb. .80
Casein, absolutely chem. pure. ⎭ ⎩ ..	lb. 2.50
Caseins, vegetable, see Conglutin, and Legumin	
Cassius's Purple, see Gold, Tin-precipit. of	
Catechin (Catechuin), [Catechuic Acid].....	oz. 2.00
Catechol, see Pyro-catechin..............	
Cathartin, in Extract-form,—(*not* identical with Cathartic Acid, – which *see also!*).....	oz. 5.00
Caustic, lunar, see Silver, nitrate, molded..	
" mitigated (toughened), see Silver, nitrate, diluted: *U. S. Ph.*,—and others	

☞ When ordering, specify: "**MERCK'S**"!

	Containers incl.		
Caustic, Filhos's, (Fused Vienna Caustic), see Potassium, hydroxide, with Lime, [4:1], fused......................			
" Vienna, powder, see Potassium, hydroxide, with Lime, [2:1], powder:......			
Cedrin, cryst.; from Cedron-seed.—Transparent crystals; wholly volatilizable; readily soluble in Water. —(Febrifuge, etc.; antidote in hydrophobia, etc.)..................	15 gr. 8.00		
Cerebrin.—Physiological preparation from brain-substance..........................	15 gr. 2.00		
Cerium, metallic, fused....................	15 gr. 7.50		
" acetate.....................................	oz. 1.00		
" bromide.....................................	oz. 1.00		
" carbonate...................................	oz. .75		
" chloride.....................................	oz. .35		
" lactate	oz. 2.00		
" malate	oz. 3.50		
" nitrate.....................................	oz. .40		
" oxalate—*U. S. Ph.*—of Sesqui-oxide...	oz. .15		
" oxide (per-oxide), pure..................	oz. 1.00		
" sulphate (bi-sulphate) of Per-oxide....	oz. .35		
" sulphate of Sesqui-oxide................	oz. .40		
Cetrarin (Cetraric Acid)...................	15 gr. 1.00		
Chalk, prepared (levigated),—*Creta præparata, U. S. Ph.*, — [Purified (elutriated) Carbonate of Calcium]	lb. .12		
Chameleon Mineral, (Mineral Chameleon), see Potassium, manganate			
Charcoal, animal (Bone-), [Bone-black], (Carbo Ossium; Spodium),—purified, — wet process;—[so-called "Ivory Black"—Ebur ustum];— *Carbo animalis purificatus, U. S. Ph*...........	lb. .50		
" do., (do.), pure,-wet process,-[do.- etc.]	lb. 1.25		
" Meat-, (Carbo Carnis), [Medicinal Animal Charcoal,—for *internal* use], pure	lb. 3.00		
" Blood-, (Carbo Sanguinis).............	lb. 2.00		
" " purified by acid	lb. 2.25		
" Sponge-, (Burnt Sponge—Spongia usta [tosta]; Carbo Spongiæ), powder....	lb. .75		
Chelerythrine (Chelerythria)...............	15 gr. 1.25		
Chelidonine (Chelidonia), pure	15 gr. 1.00		
" hydrochlorate	15 gr. 1.00		
" sulphate	15 gr. 1.00		
Chinidine, Chinine (Chinia), Chinium, Chinoidine ("*Chinoidinum*" *of U. S. Ph.*), Chino - iodine, Chinoline (Chinoleine), Chinone; and compounds of these;—see Quinidine, Quinine, Quinium, Quinoidine, Quino-iodine, Quinoline, Quinone,—etc.....			
Chinoyl, see Quinone.....................			
Chitin,—from beetles.....................	15 gr. 2.50		
Chloral, — *so called by the U. S. Ph.*,—see Chloral Hydrate.........................			
Chloral, alcoholate, anhydrous............	oz. .30		
" camphorated	oz. 1.00		
Chloral Hydrate, (the so-called "*Chloral*" of the *U. S. Ph.*):			
crusts	lb. 1.50		
loose crystals...........................	lb. 1.55		
true Liebreich...........................	lb. 2.25		
according to Liebreich	lb. 2.00		
Chloral Hydrocyanate (Cyanhydrate), cryst. —Rhombic-prisms; white, translucent; wholly volatilizable; readily soluble in Water, Alcohol, or Ether.— [Very stable compound, acting physiologically like *Prussic Acid*; hence, a desirable substitute for Bitter-almond and Cherry-laurel Waters.]	½ oz. vls. oz. 2.00		

☞ When ordering, specify: "**MERCK'S**"!

	Containers incl.
Chloral, meta-	oz. 1.00
Chlor-anile	15 gr. .30
Chlorine Bromide, (*Chlorum bromatum*), so-called, see Bromine, chloride	
Chlorine-water (Solution of Chlorine in distilled Water)	
Chloro-ethyl (Mono-chlor-ethane), chlorinated compounds of, see Ether, hydrochloric, etc.	
Chloroform (Ethyl chloroform), pure,—*Chloroformum purificatum, U. S. Ph.,*—conforming to Ph. G. II	lb. 1.50
" from Chloral	lb. 3.00
" English (British),—in original jars	lb. 2.50
" chem. pure—according to British standard, (purissimum uso anglico)	lb. 1.50
Chloroform. Methyl-, see Methyl Chloroform	
Chlorogenine (Alstonine),—from Alstonia-bark—[Alstonia constricta Apocyneæ]	15 gr. 1.75
Chlorophyll, chem. pure	15 gr. .60
" technically pure,—for use in the arts; free from Cupric Oxide	oz. .50
Chole-stearin (Cholesterin)	15 gr. .50
Chole-stearin Fat, see Lanolin	
Chondrin (Cartilage Gelatin)	15 gr. .50
Chrom-aci-chloride (Chromyl Di-chloride), see Chromium, di-oxy-di-chloride	
Chrome Alum, see Alum, chromic	
Chromium (*Chrome*), metallic, fused	15 gr. 1.00
" acetate	oz. .50
" chloride, sesqui-, see Chromium, sesqui-chloride	
" di-oxy-di-chloride, (Chrom-aci-chloride, Chromyl Di-chloride), [Chloro-chromic Anhydride]	
" fluoride	oz. .50
" hydroxide, Chromic, see Chromium, oxide, hydrated	
" nitrate	oz. .35
" oxalate	oz. .50
" oxide (sesqui-oxide), [Chromic oxide], anhydrous, dry	lb. 1.00
" " do., chem. pure	lb. 1.25
" " hydrated, (Chromic Hydroxide), dry	lb. .75
" oxy-chloride, see Chromium, di-oxy-di-chloride	
" phosphate, cryst.	oz. 1.75
" sesqui-chloride	oz. 1.50
" sulphate	oz. .30
Chromium and Potassium, sulphate, see Alum, chromic	
Chromyl Di-chloride, see Chromium, di-oxy-di-chloride	
Chrys-aniline (so-called "Phosphine"), see under Aniline and Phenol Dyes: Yellow	
Chrys-arobin,—*U. S. Ph.* and *Ph. G. II*,—(so-called "Medicinal Chrysophanic Acid")	oz. .40
N. B.—*True* Chrysophanic (Rheic) Acid, see Rhubarb constituents: Rhein.	
Chrysoidine, cryst., see under Aniline and Phenol Dyes	
Cicutine (Conicine), see Coniine	
Cimicifugin.—Resinoid from Black Snake-root, (Black Cohosh), [Cimicifuga (Actæa) racemosa]	oz. 2.00
Cinchonidine (Cinchonidia) [*Alpha*-Quinidine] (Cinchovatine), pure, cryst.	oz. .60
" borate	oz. .75
" hydrochlorate	oz. .60
" salicylate	oz. .50

☞ When ordering, specify: "MERCK'S"!

	Containers incl.			
Cinchonidine — (*as above!*), — **sulphate. Zimmer's**; conforming to *U. S. Ph.*	1 oz.vial 5 oz.tin,oz.			
" tannate	oz. .50			
" tartrate	oz. .75			
Cinchonine (Cinchonia), chem. pure, cryst., — *U. S. Ph.*, —free from Cinchotine	oz. 1.50			
" pure, cryst.	oz. .35			
" " precipitated	oz. .28			
" benzoate	oz. 1.00			
" ferri-citrate, [25% Cinchonine]	oz. .30			
" hydrochlorate	oz. .23			
" salicylate	oz. .40			
" sulphate,— *U. S. Ph.*,— large cryst.	oz. .24			
" tannate	oz. .30			
Cinchovatine, see Cinchonidine				
N. B.—Other *Cinchona derivatives*, see Quinidine, Quinine, etc.; also: Acid, quinic, etc.				
Cinis stanni (Jovis), [Tin Ash], see Tin, oxide, grey				
Cinnabar, artificial, best, see Mercury, sulphide, red, *U. S. Ph.*				
Cinnam-alcohol, see Styrone				
Cinnamene (Cinnamol), see Styrol				
Cinnyl (Styryl) **Cinnamate**, see Styracin				
Citrullin, see **Colocynthidin**				
Coal-tar Benzol, ⎰ (so-called "Coal-tar Benzin"), " **Naphtha** ⎱ —see *Benzene*, anthracic				
" **Dyes** (*Colors*), see **Aniline and Phenol Dyes**.				
Cobalt, metallic, [98–99%], granulated	oz. .50			
" " pure	oz. 2.00			
" acetate	oz. .70			
" ammonio-sulphate, see Cobalt and Ammonium, sulphate,—(below!)				
" arseniate (arsenate)	oz. .65			
" " technical, see under Cobalt, oxide				
" carbonate, pure	oz. .50			
" " technical, see under Cobalt, oxide				
" chloride, pure, cryst.	oz. .45			
" chromate	oz. .65			
" cyanide	oz. 1.00			
" nitrate, pure, cryst.	oz. .30			
" " —solution	oz. .25			
" oxalate, pure	oz. .50			
" oxide, chem. pure	oz. 1.00			
" " *for the Porcelain manufacture* and other technical uses:—				
blue, F. U.	oz. 1.25			
black, I a, F. F. K. O.	oz. 1.00			
grey, II a, F. K. O.	oz. .75			
black, III a, R. K. O.	oz. .75			
" IV a, P. O.	oz. .75			
—arseniate,—A. K. O.	oz. .70			
—carbonate,—K. O. H.	oz. .75			
—phosphate,—P. K. O.	oz. .85			
" phosphate	oz. .50			
" " technical, see under Cobalt, oxide				
" sulphate, pure, cryst.	oz. .25			
" tartrate	oz. .75			
Cobalt and Ammonium, sulphate	oz. .35			
" and **Potassium**, cyanide, see Potassium, cobalti-cyanide				
Cobaltum Mineral, so-called,— (so-called "Metallic" Arsenic),—see Arsenic, cryst.				
Coca-ethyline	15 gr. 3.00			
Cocaine Merck:				
pure	15 gr. .75			
" synthetically prepared	15 gr. 7.00			
benzoate	15 gr. .75			
carbolate, see Cocaine, phenate				
borate	15 gr. .75			

☞ When ordering, specify: "**MERCK'S**"!

	Containers incl.

Cocaine Merck,—*continued:*
- citrate .... 15 gr. .75
- hydrobromate .... 15 gr. .75
- hydrochlorate, chem. pure, cryst., perf. white .. 15 gr. .45
- nitrate .... 15 gr. .75
- oleate [5% of Alkaloid] .... ½ oz.vls.oz. 3.00
- " [10% "] .... ½ oz.vls.oz. 4.00
- " [50% "] .... ½ oz.vls.oz. 12.00
- phenate (phenylate, carbolate), [Phenol-Cocaine],—soft extract consistency .... 15 gr. 1.00
- phtalate,—syrupy consistency.—Very easily soluble in Water and in Alcohol .... 15 gr. 1.00
- salicylate .... 15 gr. .75
- sulphate .... 15 gr. .75
- tannate .... 15 gr. .75
- tartrate .... 15 gr. .75

☞ *N. B.* — **These Cocaines** *bear in absolute perfection* ALL TESTS,—*including the one by Ammonia, recently recommended by* MACLAGAN, *and the Intensified Permanganate test* (*see* MERCK'S BULLETIN, *No. 2 of Vol. 1*).

Cocaine Discs,—in tubes of 100 ....

Codeine (Codeia), pure, cryst.,—*U. S. Ph.*... ½ oz.vls.oz. 4.75
- " acetate .... ½ oz.vls.oz. 12.00
- " citrate .... ½ oz.vls.oz. 11.50
- " hydrobromate .... ½ oz.vls.oz. 10.00
- " hydrochlorate .... ½ oz.vls.oz. 6.00
- " hydro-iodate (hydriodate) .... ½ oz.vls.oz. 10.00
- " nitrate .... ½ oz.vls.oz. 12.00
- " **phosphate, soluble, Merck,**—soluble in 4 parts Water .... ½ oz.vls.oz. 9.00
- " salicylate .... ½ oz.vls.oz. 12.00
- " sulphate,—soluble in 35-40 parts Water ½ oz.vls.oz. 4.50
- " valerianate .... ½ oz.vls.oz. 12.00

Codeine and Morphine, hydrochlorate, see Salt, Gregory's ....

Coffeine, see Caffeine ....

Colchicein .... 15 gr. 2.50

Colchicine Merck, chem. pure, cryst. ...... 15 gr. .50
- " pure, powder ....
- " tannate .... 15 gr. .45

Colcothar, pure, see Iron, oxide, red, *anhydr.*

Collections (Specimen Cases) of ⎫ see *Speci-*
  Alkaloids, Glucosides, etc. ⎬ *men Collec-*
- " of Metals .... ⎪ *tions,* — at
- " of Physiological Preparations ⎭ End of List.

Collodion, simple, [2% Pyroxylin] .... lb. 1.20
- " *U. S. Ph.*,-double, [4% "], Ph. G. II lb. 1.25
- " Ph.Belg.new, " [4% "], flexible .. lb. 1.30
- " triple .... [6% "] .... lb. 1.35
- " cantharidal (vesicatory),—Ph. G. II .... lb. 2.50
- " flexible (elastic) .... lb. 1.25
- " iodized .... lb. 2.50
- " iodoformized .... lb. 4.00

Collodion Cotton,— Ph. G. II, —(Soluble Gun Cotton, Pyro-xylin, Collo-xylin, Cotton Xyloidin).—*Can be shipped only when wet*... oz. .40

Colocynthidin (Citrullin) .... 15 gr. .75

Colocynthin, chem. pure .... 15 gr. .75

Columbin .... 15 gr. 1.25

Conchinine, see Quinidine ....

Condurangin.—Glucoside from Condurango-bark ....

Conessine, pure, cryst. .... 15 gr. 6.00

Conglutin (*Vegetable Casein* from almonds) oz. 1.50

Congo Paper,— according to Prof. Riegel.— (Test-paper for Hydrochloric Acid in the stomach.) .... quire .75

Congo Red, see under Aniline and Phenol Dyes: Red ....

☞ When ordering, specify: "**MERCK'S**"!

MERCK'S INDEX.

	Containers incl.			
Coniferin	oz. 3.50			
Coniine Merck, (Conicine, Cicutine), pure	⅛ oz.vls.oz. 6.00			
" hydrobromate, cryst.	15 gr. .50			
" " powder	15 gr. .50			
" hydrochlorate	15 gr. .75			
Convallamarin	15 gr. .75			
Convallarin	15 gr. .60			
Convolvulin (White Resin of True Jalap).— The pure *Glucoside* from the True Jalap-root—from Ipomœa purga				
N.B.—*See, also:* Resins: Jalap,—brown, fr. the *true* Root;—and, do., etc., Ph. G. II.				
Copaiva, see Balsams: Copaiva				
Copper (Cuprum), double and triple salts of, see "Copper and —" (below!)				
" metallic, granulated	lb. .75			
" " in scales	lb. 1.50			
" " filings	oz. .25			
" " shavings	lb. .75			
" " reduced, powder	oz. .25			
" acetate, basic, (sub-acetate), refin'd, powder; [Purified Verdigris—Ærugo purificata], (Viride æris purific.)	lb. .75			
" " normal (neutral), pure, cryst.,— *U. S. Ph.;*—[Crystallized Verdigris—Ærugo destillata (crystallisata)], (Flores virides æris)	lb. 1.00			
" albuminate	oz. .50			
" aluminated, (so-called "Divine Stone," or "Ophthalmic Stone"; also called "Copper-alum"), in plates	lb. .60			
" " in pencils	lb. 1.00			
" " powder	lb. .60			
" ammoniated, so-called, see Copper and Ammonium, sulphate				
" arseniate (arsenate)	oz. .30			
" arsenite	oz. .30			
" benzoate	oz. .50			
" bi-chloride, pure	lb. .80			
" " cryst., commercial	lb. .50			
" borate	oz. .30			
" bromide	oz. .60			
" butyrate	oz. .80			
" carbonate, green (di-cupric) Artificial Malachite (Mountain-green)	lb. .75			
" " " chem. pure	lb. 1.00			
" " blue (sesqui-cupric), [Artificial *Blue* Malachite, (Mountain-blue); Verditer],—A1 English	lb. 1.00			
" chlorate	oz. .85			
" chloride (mono-chloride), white	lb. 2.50			
" " bi-, see Copper, bi-chloride				
" chromate	oz. .25			
" " liquid	lb. .85			
" citrate	oz. .40			
" cyanide	oz. .35			
" ferro-cyanide, see Cop. and Iron, cyanide				
" formate, cryst.	oz. .70			
" iodide	oz. .75			
" lactate	oz. .50			
" nitrate, cryst., commercial	lb. .60			
" " " pure	lb. .70			
" " " chem. pure	lb. .75			
" nitro-prusside (nitro-prussiate; nitro-ferri-cyanide)	oz. 1.50			
" oleate	oz. .25			
" oxalate	lb. 1.85			
" oxide, black (Cupric), [mon-oxide], pure, powder	lb. .90			
" " " pure, coarse granul. for analyses	lb. 1.75			
" " " " wire	lb. 2.00			

☞ When ordering, specify: "MERCK'S"!

	Containers incl.			
Copper, oxide, black,-(*as above!*),-technical	lb. .40			
" " " hydrated, pure	oz. .50			
" oxide, red (Cuprous), [sub-oxide], pure	lb. 1.50			
" " " commercial	lb. .60			
" phosphate	oz. .25			
" phosphide (phosphuret), powder	oz. .50			
" rhodanide, see Copper, sulpho-cyanate.				
" salicylate, powder	oz. 1.00			
" " in sticks	oz. 1.50			
" sub-acetate,(Purified Verdigris), see Copper, acetate, basic				
" sulphate, basic (tetra-cupric)	lb. 1.75			
" " neutral, (Copper Vitriol; Blue Vitriol), ch. pure,— *U. S. Ph.*	lb. .40			
" " " molded (fused), in sticks	lb. 1.00			
" " " caustic pencils, turned	doz. 1.00			
" " " " " mounted in wood	doz. 3.50			
" " " cryst., commercial	lb. .30			
" sulphide (sulphuret), fused	lb. 1.10			
" " granulated	lb. 1.10			
" " powder	lb. 1.10			
" " —by wet process	lb. 2.00			
" sulpho-carbolate (phenol-sulphonate, sulpho-phenate), chem. pure	oz. .35			
" sulpho-cyanate(thio-cyanate;rhodanide)	oz. .30			
" tannate	oz. .25			
" tartrate	oz. .30			
" thio-cyanate, see Cop., sulpho-cyanate.				
Copper and **Ammonium**, acetate ⎫	oz. .35			
" and do., chloride Ammoni-	oz. .25			
" " " chromate ⎬ cupric	oz. .40			
" " " cyanide salts.	oz. 1.00			
" " " nitrate ⎭	oz. .30			
" " " sulphate, (Ammonio-sulphate of Copper; so-called "Ammoniated Copper")	lb. .80			
" and **Calcium**, acetate, cryst.	oz. 1.00			
" and **Iron**, cyanide,(*Cupric Ferro-cyanide*)	lb. 2.50			
" and **Platinum**, double *and triple* salts, see under Platinum double Cyanides; and, do. *triple* Cyanides				
" and **Potassium**, chlorate	lb. 2.50			
" " " chloride	lb. .75			
" " " cyanide	lb. 2.50			
" and **Sodium**, chloride	lb. 1.25			
Copper, **Platinum**, and **Ammonium**, cyanide-cyanuret, see und. Platin. triple Cyanid.				
Copper **Alum**, ("Divine Stone"), so-called, see Copper, aluminated				
" **Vitriol**, (*Blue Vitriol*), see Copper, sulphate, neutral, *U. S. Ph.*; and others.				
Corallin, see under Aniline and Ph. Dyes: Red				
Corrosive Sublimate, see Mercury, bi-chloride, *U. S. Ph.*; etc.				
Corydaline, cryst.	15 gr. 2.00			
Cosin Merck, and **Coussein Merck**, see **Kosin**, and **Koussein**				
Cosmolin, see Vaselin				
Cotoin, true	15 gr. 3.00			
" para-, commercial	15 gr. .35			
" " chem. pure, free from Leucotin	15 gr. 1.00			
" **Hydro-**	15 gr. .30			
Coumarin, see Cumarin				
Cream (and **Crystals**) of **Tartar**, see Potassium, bi-tartrate. *U. S. Ph.*; and others				
" (and **Scales**) of do.: "soluble" (*so-called;*-AND: *perfectly soluble*),—[Borax-Tartar];—see Potassium and Sodium, boro-tartrate,—*and:* do. do. do., do.,— *in scales*				

☞ When ordering, specify: "MERCK'S"!

	Containers incl.			
Creasote (Creosote), **pure,** — Ph. G. II, — from *Beech*-tar	lb. 2.00			
" pure, white, true....... { —From }	lb. .59			
" chem. pure, white, true. { Coal-tar. }	lb. .85			
Creatine (Kreatine)	15 gr. 3.50			
Creatinine (Kreatinine)	15 gr. 6.00			
" with **Chloride of Zinc**	15 gr. 1.75			
Creolin.—(Antiseptic; non-toxic deodorizer, disinfectant, and anti-bacterial; *claimed to exceed Carbolic Acid* in deodorizing power, while being absolutely safe!)	lb. 1.00			
N.B.—*See, also:* **Mollin Ointments: Creolin.**				
Creosote, see **Creasote**				
Cresol, see **Acid, cresylic**				
Creta præparata, *U. S. Ph.*,—(Creta lævigata),—see **Chalk, prepared**				
Crocus (Saffron) of **Antimony,** [*Crocus metallorum*], see **Potassa, antimonio-sulphurated,** *washed*				
" of **Iron,** *aperient,* (Crocus martis aperitivus), see **Iron, oxide, brown,** [so-called sub-carbonate]				
" " " *astringent,* (Crocus martis adstringens), see **Iron, oxide, red,** *anhydrous*				
Croton-chloral Hydrate, see **Butyl-chloral Hydrate.**				
Cryptopine.—Alkaloid from Opium	15 gr. 7.00			
Cubebin	15 gr. .35			
Cumarin (Coumarin) [Cumaric Anhydride, Cumarylous Acid] (Tonka-bean Camphor).	oz. 2.50			
Cumene (Cumol), [Iso-propyl-benzene],— boiling-point 160–170° C [320–338 F]	lb. 1.00			
Cuprein,—from *Cuprea-bark,*—see **Vieirin..**				
Cuprum, and compounds, see **Copper, etc..**				
Curare (Urari, **Woorali, Woorara, Woorari**), tested for efficacy	15 gr. .25			
Curarine, chem. pure, free from Curine	15 gr. 3.00			
Curcuma Paper, see **Paper, Turmeric-.**				
Curcumin (*Curcuma Yellow,* Turmeric Yellow)	15 gr. .35			
Cyan-amide	15 gr. 2.00			
Cyanine (Quinoline Blue), [Chinoline-iodo-cyanine], chem. pure, large crystals	15 gr. 1.00			
Cyano-amyl, see **Amyl, cyanide**				
Cyano-ethyl (Cyanide of Ethyl), see **Ether, hydrocyanic**				
Cyano-methyl, see **Methyl, cyanide**				
Cyclamin, cryst.	15 gr. 1.00			
Cymene (*Cymol*), **para-,** [para-Methyl-propyl-benzene], crude,—from Camphor	¼ oz.vls.oz. 2.00			
" do.,—from Oil of Roman Cumin	⅛ oz.vls.oz. 1.50			
Cytisine, nitrate, cryst.	15 gr. 5.00			

☞ When ordering, specify: "**MERCK'S**"!

	Containers incl.			

☞ When ordering, specify: "MERCK'S"!

	Containers incl.
Daggett (Degutt), see Oils, divers: Birch, empyreumatic....................	
Dahlin (Alant-starch), see Inulin..........	
Daphnetin............................	15 gr. 5.00
Daturine, pure, cryst., (True or heavy Daturine, identical with **Atropine**);—from Datura Stramonium.....................	15 gr. 2.50
" **hydrochlorate**, pure....................	15 gr. 2.50
" **sulphate**, pure........................	15 gr. 2.50
Degutt (Daggett), see Oils, divers: Birch, empyreumatic........................	
Delphinine............................	15 gr. 1.00
Dextrin, chem. pure, precipit. by Alcohol..	lb. 1.00
" pure,—Ph. G. I.....................	lb. .75
" purest granulated, for use in the arts..	lb. .50
" white or yellowish, " " " ..	lb. .20
Dextrose (*Dextro-glucose*), see **Grape-sugar**, chem. pure......................	
Di-amido-benzene (-benzol), **meta-**, hydrochlorate,—(Hydrochlorate of meta-Phenylene-di-amine).....................	oz. 3.50
Di-amido-toluene (-toluol), see Tolylene-di-amine	
Diamond Ink, so-called,—for Glass-etching.....	oz. .50
Diastase of Malt, (Maltin)................	oz. 1.50
Di-benzoyl, see Benzile..................	
Di-chlor-ethane, **Alpha-**, see Ethylidene, chloride (bi-chloride)...............	
" **Beta-**, see Ethylene, chloride (bi-chl.).	
Di-chlor-hydrin	oz. 1.00
Di-chlor-methane, see Methylene Chloride (Bichloride) Merck, chem. pure..............	
Di-chlor-naphthalene, **Alpha-**, see Naphthalene, Alpha-di-chlorated.............	
Didym (Didymium), metallic, powder......	15 gr. 9.00
" carbonate	15 gr. 1.00
" chloride......................	15 gr. 1.00
' nitrate	15 gr. .75
" oxide	15 gr. 1.00
" sulphate	15 gr. .75
Di-ethyl-acetal, see Acetal................	
Digitalis preparations :	
Digitalein (Schmiedeberg's)	15 gr. 1.25
Digitalin Germanic Merck, pure, powder	$\frac{1}{8}$ oz vls.oz. 3.75
" pure, amorph.,—Ph. Gallic. and Ph. Belg.	15 gr. 1.50
" crystallized,—so-called,—see *Digitin*.	
" purified,—Ph. Austr. VI............	15 gr. .75
Digitin (*so-called* "Crystallized Digitalin")	15 gr. 1.25
Digitoxin, chem. pure......................	1½ gr.vial 2.00
Di-methyl-acetal, pure...................	oz. 1.50
Di-methyl-aniline, pure..................	oz. .50
Di-methyl-aniline Orange, see under Aniline and Phenol Dyes: Orange............	
Di-methyl-benzene (-benzol), see Xylene..	
Di-methyl-carbinol, see Alcohol, propylic, Iso-...........................	
Di-methyl-ketone, see Acetone	
Di-methyl-oxy-quinizine (-chinizine), see Antipyrine	
Di-methyl-pyridine, see Lutidine	
Di-nitro-benzene (-benzol, -benzide), [*Bi-*nitro-b., etc.], **meta-**, commercial...	lb. 2.00
" do., pure	
Di-nitro-naphthalene (*Bi*-nitro-naphthal.)	oz. 1.50
Di-nitro-toluene (-toluol), [*Bi*-nitro-tol.]..	lb. 3.00
Di-oxy-benzene (-benzol), **ortho-**, see Pyrocatechin	
" **meta-**, see Resorcin.................	
" **para-**, see Hydro-quinone	

☞ When ordering, specify : "**MERCK'S**"!

	Containers incl.		
Di-oxy-toluene (-toluol), **meta-, symmetric,** see Orcin............			
Di-phenyl-amine, chem. pure, cryst.......	oz. .35		
" crude...............................	lb. 1.50		
" sulphate, chem. pure.................	oz. .40		
Di-phenyl-ethylene, see Stilbene.........			
Di-phenyl-imide, see Carb-azole..........			
Di-phenyl-mercury (*not* = Mercury Di-phenate!);—see remark under *the latter!*			
Di-platos-amine, see Platos-amine, di-....			
Di-resorcin (Di-resorcinol)..................	oz. 1.25		
Discs (Gelatin Discs), medicated,—for Ophthalmology,—see under Atropine, Cocaine, Duboisine, and Physostigmine...........			
Ditaine, cryst...........................	15 gr. 3.50		
" sulphate..............................	15 gr. 3.50		
Divine Stone, so-called, see Copper, aluminated........................			
Donovan's Solution, see Solutions: Arsenic and Mercury Iodides, *U. S. Ph.*........			
Duboisine (Duboisia-Alkaloid), pure, amorphous...........................			
" pure, cryst...........................	15 gr. 4.00		
" hydrochlorate........................			
" sulphate, amorphous..................	15 gr. 1.75		
Duboisine Discs,—in tubes of 100........			
Dulcit (Dulcin, Dulcol, Dulcose, Dulcitol), see Melampyrit......................			
Dutch Drops, (Haarlem Oil), see Oils, divers: sulphurated Linseed-, terebinthinated.............................			
Dutch Liquid, see Ethylene, chloride (bichloride)....			
Dyers' Salt, (Pink Salt), see Tin and Ammonium, chloride...................			
Dyslysin...............................	15 gr. .75		
Dzondi's Solution, caustic ammoniacal, see Ammonia, Spirit of.....................			

☞ **When ordering, specify: "MERCK'S"!**

	Containers incl.			
"**E**au des Carmes," see Spirit, Balm,—compound...........................				
Ebur ustum, see Charcoal, animal, purified, U. S. Ph.; and, pure.....................				
Ecgonine............................	15 gr. 5.00			
Egg preparations,—all soluble: Albumen, dried, in scales.—(Its solution in Water replaces fresh Egg Albumen for all dietetic or technical uses.)...........				
Albumin................................ ⎫				
" I, inodorous................... ⎬ see Albumin				
" in scales,—free from Fibrinous matter;—for laboratories.....				
" impalpable powder;—for gilders, stampers, etc. .... ⎭				
—(See, at same place, also *other kinds* of Albumin,—from blood, etc.).				
Yelk (Yolk), [Vitellus ovi], dried,—sifted; —for bird-food..................				
" dried,—light, flocculent powder;—for human food.....................				
" do.,—in spongious flakes;—for human food, and for rearing exotic birds..				
Elaidin...............................	15 gr. .75			
Elastin, dry............................	15 gr. .50			
Elaterin Merck, cryst.—(Elateric Anhydride)..	15 gr. 1.50			
Elaterium—(*sediment* of the fruit-juice of Ecballium elaterium—Squirting Cucumber)—[*Elaterium Clutterbuck*]....	½ oz.vls.oz. 2.75			
" black, true, (Elaterium nigrum verum), —[*inspissated* fruit-juice of above-named plant],—see Extracts: Squirting Cucumber; aqueous				
Elayl, etc., see Ethylene, etc..............				
Elecampane-camphor, solid, see **Helenin**........				
" **liquid,** see Alantol..................				
Emetine (Emetia).—*Alcoholic Extract* of Ipecacuanha-root.......................	oz. 3.00			
" **chem. pure,** light-colored.—*The Alkaloid* of Ipecacuanha-root...............	15 gr. 1.50			
Emplastrum, see Plaster..................				
Emulsin...............................	15 gr. .35			
Eosin, see under Aniline and Phenol Dyes: Red................................				
Ephedrine, hydrochlorate, cryst. — (A mydriatic.)	15 gr. 3.00			
Epsom Salt, see Magnesium, sulphate, (*etc.*)				
Erbium, metallic........................	15 gr. 7.50			
oxide	15 gr. 1.50			
Ergotin (Ergotinum), *so called* by Ph. G. II; see Extracts: Ergot of Rye,—*Ph. G. II.*				
" Bonjean	oz. .36			
" " purified,—for injections..........	oz. .50			
" " dry, with Sugar of milk	oz. .50			
" Wernich, dialyzed, pure, liquid	oz. 1.50			
" " " " inspissated....	oz. 1.75			
" " " " dry..........	oz. 2.50			
" Wiggers, pure, dry....................	oz. 6.00			
" d'Yvon.............................	oz. .75			
" Bombelon, liquid.....................	oz. 2.25			
" " inspissated.....................	oz. 2.25			
" " dry............................	oz. 2.50			
" Denzel............................	oz. 1.75			
" Kohlmann, liquid....................	oz. .50			
Erythrit (Erythrol, Erythro-mannit, Erythro-glucin)............................	15 gr. .50			
Erythrophleine, hydrochlorate, — from Sassy-bark, (Mancona-bark).—[Ophthalmological local anesthetic.].......................	15 gr. 4.00			

☞ **When ordering, specify: "MERCK'S"!**

	Containers incl.		
Erythro-retin, see under Rhubarb constit.			
Esculin	15 gr. .50		
Eserine, see **Physostigmine**			
Eserine Discs;—Gelatine;—Paper;—see Physostigmine Discs; etc.; etc.			
Essence of Mirbane,—*so-called,*—see Nitro-benzene			
" of Niobe,—*so-called,*—see Methyl, benzoate			
" —*so-called*—of Whey, see Rennet Wine			
Essences,—*real!*—see Essential Spirits			
Essential Oils — (are inserted in alphabetical place of : *Oils, Essential*)—see, *after:* "Oils, divers."			
Essential Spirits, (Essences):			
Arrack			
Cognac, brown			
Curaçao (Curaçoa)			
French Brandy, white			
Grape-marc			
Muscat-Lunel			
Prunes,—(Slibowitz)			
Rum Aroma			
Rum, finest Jamaica			
" " Kingston			
" —concentrated;-(so-called "Rum-oil")			
" white			
Slibowitz, see Essential Spirit, Prunes			
Whiskey (Grain-spirit),—["Korn-Essenz"]			
Wild sour Cherry, ("Weichsel")			
N.B.—See also *Fruit and Flavoring Ethers:* Rum; and, Rye.			
Ester, aceto-acetic, see Ethyl, aceto-acetate			
N. B.—*Other Esters* (Acid-and-Hydrocarbon-Hydroxyl compound Ethers)—[Salts of Alcohols ; Organo-base Salts],—see under Ether.			
Ethal (Cetylic Alcohol), chem. pure	oz. 1.50		
Ethene, etc., see Ethylene, etc.			
Ether, acetic, (Acetate of Ethyl), [Vinegar Naphtha],— sp. gr. 0.902,—Ph. G. II	lb. 2.50		
" " twice rectified,— sp. gr. 0.890,— U. S. Ph.	lb. 2.25		
" " rectified,— sp.gr. 0.870-0.880	lb. 2.00		
" aceto-acetic, (Aceto-acetic Ester), see Ethyl, aceto-acetate			
" amylic	oz. 2.00		
" amylo-acetic, etc., see Amyl, acetate, etc.			
" " -nitrous, etc., see Amyl, nitrite, U. S. Ph.; and others			
" anesthetic, Wiggers's, see Ether, hydrochloric, poly-chlorated			
" benzoic, (Benzoate of Ethyl), pure, from true Benzoic Acid	lb. 6.50		
" " from artificial Benzoic Acid	lb. 3.50		
" butyric, (Butyrate of Ethyl)	lb. 3.75		
" " absolute	lb. 6.00		
" " concentrated, best	lb. 4.00		
" cantharidated,—Ph. G. II	lb. 4.00		
" carbolic (*ethylo*-carbolic), Carbolate of Ethyl), see Phenetol			
" cinnamylo-cinnamic, see Styracin			
" —*so-called,*—cocoinic(cocinic), [so-called "Cocoa-ether" or "*Cognac* Ether"]	oz. .75		
" ethylic, see Ether, sulphuric, so-called, *U. S. Ph.s;* etc.			
" ethylo-*phenic* (-*carbolic*), see Phenetol			

☞ **When ordering, specify: "MERCK'S"!**

	Containers incl.			
Ether, formic, (Formate of Ethyl)	lb. 1.95			
" " concentrated	lb. 2.00			
" " absolute	lb. 3.50			
" glycerino-salicylic, (Glycerin Salicylate)	oz. 2.00			
" **hydrobromic, Merck, chem. pure, (Bromide of Ethyl**; Mono-brom-ethane). — [An anesthetic, safer and milder than Chloroform, and especially adapted for *small* operations.]	oz. .40			
" hydrochloric, poly - chlorated, (Poly-chlorated Chloride of Ethyl; Wiggers's Anesthetic Ether),— sp. gr. 1.50	oz. 1.00			
" " mono-chlorated, see Ethylidene, chloride (bi-chloride)				
" hydrocyanic, (Cyanide of Ethyl)				
" **hydro-iodic** (hydriodic), **[Iodide of Ethyl**; Mono-iod-ethane]	oz. .80			
" methylo-acetic, see Methyl, acetate				
" methylo-phenic, see Anisol				
N. B.— *Other compound Methyl-*ethers, see under Methyl.				
" muriatic, etc., see Ether, hydrochloric, etc.				
" naphthylo-salicylic, Beta-, see **Betol**				
" nitrous, true, (Nitrite of Ethyl),—[15%]	lb. 2.50			
" oenanthic (œnanthic), finest *limpid*	Grape- or so-called Cognac Oil			
" " rectified, finest *colorless*				
" " natural green				
" " artificial				
" oxalic, (Oxalate of Ethyl), pure	oz. .75			
" pelargonic, (Pelargonate of Ethyl)	oz. .60			
" —*so-called,*—petroleic; (*Petroleum Ether*); —*Benzinum, U. S. Ph.;*—see Benzin, petroleic, boil.-pt. 50–60° C				
" phenol-ethylic (*ethylo-phenic*), [Phenate of Ethyl], see Phenetol				
" phenylo-salicylic, see Salol				
" —*so-called,*—pyro-acetic; see Acetone				
" —*so-called,*—saccharic; (*not* Saccharate of Ethyl; but the so-called "*Sugar-Ether*"!)				
" salicylic, (Salicylate of Ethyl)	oz. .75			
" sebacylic, (Sebacylate of Ethyl)	oz. 1.25			
" succinic, (Succinate of Ethyl, [Di-ethyl Succinate])	oz. 1.00			
" sulphuric (vitriolic), so-called,—[Ethylic ether; Oxide of Ethyl], (so-called "Vitriolic Naphtha"),— sp. gr. 0.730–733	lb. 1.00			
" " sp. gr. 0.725–0.728, conforming to Ph. G. II	lb. 1.05			
" " " 0.722,—*Æther fortior, U. S. Ph.*	lb. 1.10			
" " " 0.750, [74% Ethyl Oxide, 26% Ethylic Alcohol],—*Æther, U. S. Ph.*				
" tri-chlor-acetic, (Tri-chlor-acetate of Ethyl)	oz. 1.50			
" valerianic (*iso*-valerianic), [*Iso*-valerianate of Ethyl]	oz. .65			
" vitriolic, so-called,–(Ethylic ether),–see Ether, sulphuric, so-c., *U. S. Ph. s;* etc.				
" Wiggers's anesthetic, see Ether, hydrochloric, poly-chlorated				
Ethers, Fruit and Flavoring, see Fruit and Flavoring Ethers, etc.				
Ethidene, see Ethylidene				
Ethiops, antimonial, see Mercury, antimonio-sulphide				

☞ **When ordering, specify: "MERCK'S"!**

	Containers incl.		
Ethiops, Iron-, see Iron, oxide, black.....			
" mercurial, (*Ethiops Mineral*), see Mercury, sulphide, black, — *so-called*			
Eth-oxy-Caffeine, see Ethyl-oxy-Caffeine .			
Ethyl, — acetate ; etc., etc., — see Ether, — acetic; etc., etc.....................			
" aceto-acetate, (Ethylic Ether of Aceto-acetic Acid; Aceto-acetic Ester), [Ethyl-di-acetic Acid]....................			
" bromide, see Ether, hydrobromic			
" carbolate, see Phenetol...............			
" chloride, etc., see Ether, hydrochloric, etc.			
" cyanide, see Ether, hydrocyanic.......			
" hydrosulphide (sulphydrate), see Mercaptan................................			
" iodide, see Ether, hydro-iodic...........			
" oxide, see Ether, sulphuric, so-called..			
" phenate (phenylate), see Phenetol.....			
N. B. — *Other combinations* of Ethyl, (Ethylic Acid - Esters, Halogen-Ethyls, etc.), see under Ether.			
Ethyl, Sodio- (*Natrio-*), see Sodium, ethylate			
Ethyl-amine (Amido-ethane), pure, — 33½-% solution...........................	oz. 2.50		
" chloride...........................	oz. 3.50		
" iodide............................	oz. 4.50		
Ethyl-carbinol, see Alcohol, propylic.....			
Ethyl-oxy-Caffeine (Eth-oxy-Caffeine) ...	15 gr. .50		
Ethyl-phenol, see Phenetol			
Ethylene (Ethene, Elayl), bromide........	oz. .75		
" chloride (bi-chloride), [Dutch Liquid], (Beta-Di-chlor-ethane)..............	oz. .65		
" iodide, cryst.	oz. 2.50		
Ethylene-glycol (Ethylene Alcohol)	oz. 5.00		
Ethylidene (Ethidene), chloride [bi-chloride]; (Mono-chlorated Hydrochloric Ether, Mono-chlorated Ethyl Chloride), [Alpha-Di-chlor-ethane]	oz. 1.00		
Eucalyptol (Rectified and purified Oil of Eucalyptus globulus)..............	oz. .40		
Eucalyptol, chem. pure, — acc. to Wallach; — perfectly limpid, crystallizable, — b.-p. 175-177°C [347-350.5 F], — sp. gr. 0.925; — obtained *from common Eucalyptol* by chemical re-purification	oz. 1.00		
Eugenol (Eugenic Acid; formerly called also: "Caryophyllic Acid"), — the principal constituent of Oil of Cloves; — boil.-pt. 247° C [476.6 F]......................	oz. .50		
Euonymin ⎰ Ameri- ⎰ brown.. ⎱ *Resinoids.* (*Evonymin*), ⎱ can, ⎱ green .. ⎰	oz. 1.50 oz. .90		
Euonymin (*Evonymin*) Merck, pure; — a highly pure *Resinoid* of peculiarly excellent and reliable efficacy..........................	15 gr. .50		
N.B. — *All these* — Resinoid! — *Euonymins* (or Evonymins) should not be confounded with the crystallized *Glucoside* "Evonymin," discovered by H. Meyer, which has the same toxical effect as the Digitalis Alkaloids.			
Eupione (*Crude* Pentane [Amyl Hydride])..	15 gr. .35		
Evonymit, see Melampyrit			
Excretin.................................			
Extract, — *so-called*, — Goulard's; (Vinegar of Lead); — see Solutions: Lead acetate, basic, *U. S. Ph.*			
Extracts — (See, also: *Fluid Extracts*, — after: "Extracts"!): —			
Absinthium, see Extract, Wormwood......			
Achillea (Millefolium), see Extract, Yarrow			

When ordering, specify: "MERCK'S"!

MERCK'S INDEX. 49

	Containers incl.			
Extracts,—*continued:* —[*Fluid Extracts,* see pages 61–63!]—				
Aconite: dried leaves........aqueous, soft	lb. 2.00			
" fresh " ......from juice, "	lb. 2.00			
" " " ....alcoholic, "	lb. 3.00			
" dried " —green; " "	lb. 3.00			
" recently dried leaves; " "				
Aconite: root,—Ph. G. II & Au....alco., soft	lb. 3.00			
" do.,—with powdered Licorice-root,— Ph. G. II,—[containing 50% of the soft extract].........alcoholic, dry	lb. 3.50			
Actæa (A. racemosa), see Extract, Black Cohosh................................				
Alant-root, see Extract, Elecampane......				
Alder Buckthorn, (European Buckthorn), see Extract, Frangula.................				
Alkanet (Alkanna), soft, see Alkannin......				
Aloes, Barbadoes,—Ph. Brit......aqu., dry	lb. 1.00			
Aloes, Cape,—Ph. G. II..........aqu., dry	lb. 1.00			
" " —Ph. G. I: acido sulfurico correctum sicc.;—acidulous, dry	oz. .25			
Anemone, Meadow, European, see Pulsatilla				
Angelica, European: root ....alco., soft	lb. 2.00			
" " " ....aqu., "	lb. 1.75			
Anthemis, see Extract, Chamomile, Roman				
Apple, ferrated, (Crude Malate of Iron),— Extractum ferri pomatum, Ph. G. II,— [Extractum pomorum ferratum; also called "Extractum malatis ferri"]......	lb. .65			
Arctostaphylos, see Extract, Bearberry-leaves.................................				
Arnica: flowers..................aqu., soft	lb. 1.50			
" " ....alco., "	lb. 3.50			
Arnica: root................alco,, soft	lb. 5.00			
Artemisia absinthium, see Extract, Wormwood.................................				
Artemisia maritima, see Extract, Levant Wormseed...........................				
Artemisia vulgaris, see Extract, Mugwort..				
Aspidium, see Extract, Male Fern.........				
Ava, see Extract, Kava-kava..............				
Bael, Indian, (Bengal Quince): fruit; alco., soft	lb. 3.00			
" " " " " aqu., "	lb. 2.50			
Bardane, see Extract, Burdock				
Bean of St. Ignatius, see Extract, Ignatia..				
Bearberry (*not Barberry!*) [Uva ursi]: leaves; [aqu., soft	lb. 1.50			
" do.....................alco., "	lb. 1.75			
Belladonna: dry herb............aqu., soft	lb. 1.40			
" fresh herb ....from juice, "	lb. 1.50			
" " " —with Dextrin, [50% of soft].....from juice, dry	lb. 2.50			
" " " —without admixt,.fr." "	lb. 3.00			
" " " —Ph. G. II & Neerl; alc., soft	lb. 2.50			
" " " —w. Licorice-root,-Ph. G. II, —[50% of soft],—alco., dry	lb. 3.50			
" dry herb,—green.......... " soft	lb. 3.00			
Belladonna: root.............alco., soft	lb. 2.50			
Bengal Quince, see Extract, Bael, Indian..				
Bitter Apple, see Extract, Colocynth				
Bitter Ash, see Extract, Quassia-wood.....				
Bitter Milkwort, (European Bitter Polygala), see Extract, Polygala amara				
Bitter Orange: peel (*flavedo*—that is: only the outer rind, freed from the parenchymous inner layer),—Ph. G. I; alco., soft	lb. 2.00			
do. do.: do..................aqu., "	lb. 1.75			
Bittersweet (Dulcamara): young branches; [aqu., soft	lb. 2.00			
Bitter Wood, see Extract, Quassia-wood ...				

☛ When ordering, specify: "**MERCK'S**"!

Extracts,—continued:

—[*Fluid Extracts*, see pages 61–63!]—

	Containers incl.		
Black Cohosh, (Black Snakeroot; Cimicifuga; Actæa): rhizome and rootlets	lb. 5.00		
Black Haw, (Viburnum prunifolium): bark; [alco., soft	lb. 6.50		
Black Tang .. } (Sea-wrack, Kelp-ware, Cut Weed), Bladder-wrack } [Fucus vesiculosus; Quercus marina] .... hydro-alcoholic, soft	lb. 3.75		
" —acc. to Dannecy........hydro-alco.	lb. 7.00		
Blessed Thistle, (Carduus benedictus): herb, —Ph. G. II......aqu., soft	lb. .80		
" " do............... " dry	lb. 1.25		
Bloodroot (rhizome of Sanguinaria canadensis)........................aqu., soft	lb. 2.75		
Bogbean (Menyanthes trifoliata), see Extr., Buckbean............................			
Brayera (Kousso, Cusso, Kooso): flowers; [alco., dry	oz. .90		
" do...ethereal,— (Oleoresin of Kousso)	oz. 1.00		
Bryony (Red Bryony): root......aqu., soft	lb. 1.50		
" do........................alco., "	lb. 3.00		
Buchu (Bucco): leaves ....aqu., soft	lb. 3.00		
" do........................alco., "	lb. 4.50		
Buckbean (Bogbean, Marsh Trefoil, Water Shamrock) [Menyanthes trifoliata; Trifolium fibrinum]: leaves,—Ph. G. II..aqu., [soft	lb. 1.00		
Buckthorn, Alder-(European), see Extract, Frangula			
Burdock (Lappa; Bardane): root; cold proc., [aqu., soft	lb. 1.50		
" do...................... " dry	lb. 1.75		
Cahinca (Chiococca racemosa): root..alco., [dry	oz. 1.25		
" do. ....alco., soft	oz. .75		
Calabar Bean, see Extract, Physostigma ..			
Calamus (Sweet Flag): root [rhizome],—Ph. G. II ....alco., soft	lb. 3.00		
Calendula (Garden Marigold): herb; aqu., soft	lb. 2.25		
" do........................alco., "	lb. 4.00		
Calisaya Bark, see Extract, Cinchona-bark, yellow................................			
Calumba (Columbo, Colombo): root; aqu., dry	oz. .30		
" do...................... " soft	oz. .25		
" " ....cold process, " "	oz. .40		
" " ....alco., "	oz. .50		
" " .... " dry	oz. .50		
Campeachy Wood, (Hæmatoxylon), see Extract, Logwood......................			
Cannabis indica, see Extract, Indian Hemp			
Cantharides (Spanish Flies).....ethereal,— [Oleoresin of Cantharides]	oz. 5.00		
Capsicum annuum, (Red [Pod] Pepper), [Cayenne Pepper]: fruit......aqu., soft	oz. .30		
Capsicum fastigiatum, (African [Bird] Pepper), [Guinea Pepper]: dried fruit ..ethereal, U. S. Ph.,—see Oleoresins: Capsicum			
Carduus benedictus, (Centaurea benedicta; Cnicus benedictus), see Extract, Blessed Thistle...................			
" Mariæ (marianus), [Silybum marianum], see Extract, Mary-Thistle...			
Cascara sagrada, (Chittem-bark), [Cortex Rhamni purshianæ].....hydro-alco., dry	oz. 1.00		
Cascarilla (Sweetwood): bark,—Ph. G. II, [aqu., soft	lb. 2.50		
" do...................... " dry	oz. .40		
" " ....alco., "	oz. .50		
" " .... " soft	oz. .40		

☞ **When ordering, specify: "MERCK'S"!**

	Containers incl.

Extracts,—*continued:*
— [*Fluid Extracts,* see pages 61–63 !] —
Castanea vesca, see Extract, Chestnut, European: leaves.....................

Catechu (Cutch),—from the crude extract; [aqu., dry	lb. 1.50
Celandine (Tetterwort): dry herb..aqu.,soft	lb. 1.50
" fresh flowering herb....fr. juice, soft	lb. 1.50
" fresh herb,-Ph. G. I. & Au.,-alco., "	lb. 2.75
" dry " —green.......... " "	lb. 3.00

Centaury, *European* (lesser),—[*not* a *Centaurea;*—but: Erythræa (Gentiana; Chironia) *centaurium !*):—flowering herb,—

Ph. G. I....................aqu., soft	lb. 1.50
Chamomile, German, (Matricaria) flowers; [aqu., soft	lb. 1.60
" " do.,-Ph. G. I,-alco., soft	lb. 4.00
Chamomile, Roman (English), [Anthemis]: flowers.....................aqu., soft	lb. 3.50

Chelidonium majus, see Extract, Celandine

Chestnut, European (true; sweet): leaves; [*liquid*	lb. 2.00
Chicory, Wild, (Succory): root....aqu., soft	lb. 1.40
" " do............alco., "	lb. 1.50

Chinæ cortex, see Extract, Cinchona-bark.
Chiococca racemosa, see Extract, Cahinca.

Chiretta (Chirata): flowering herb, with root; [aqu., soft	oz. .50

Chironia centaurium, see Extr., Centaury, *European*..............
Chittem-bark, see Extract, Cascara sagrada.
Christmas-rose, see Extr., Hellebore, Black
Cichorium, see Extract, Chicory..........
Cimicifuga, see Extract, Black Cohosh....
Cina (Flores Cinæ; "Semen Cinæ"), see Extract, Levant Wormseed.............

Cinchona-bark, Gray..........aqu., dry	oz. .30
" do............cold process, " soft	oz. .30
" " " " " dry	oz. .40
" " ....alco., soft	oz. .40
" " " " dry	oz. .50
" Pale...................aqu., "	oz. .40
" " " soft	oz. .35
" " ....alco., dry	oz. .60
" " " soft	oz. .55
" Red................aqu., dry	oz. 1.25
" " ....alco., "	oz. 1.15
" " " soft	oz. 1.00
" Succirubra,—Ph. G. II......aqu., "	oz. .35
" " " ....alco., dry	oz. .40
" Yellow, (True Calisaya-bark—Cortex Chinæ [Cinchonæ] regiæ); [aqu., dry	oz. .50
........cold process, " soft	oz. .75
.... " " " dry	oz. .75
" " ....alco., "	oz. .50
Coca (Erythroxylon) leaves......alco., soft	oz. .60
" do................ " dry	oz. .75

Cochlearia (Spoonwort), see Extract, Scurvy-grass.....................

Coffee: unroasted seed..........aqu., soft	oz. .50
" " " ....alco., "	oz. .50
Colchicum (Meadow-saffron) root (bulb, tuber, corm) ....alco., soft	oz. .40
" seed................... " dry	oz. .75
" root..................acetic, soft	oz. .35
" seed.................. " "	oz. .65
Colocynth (Bitter Apple): decorticated fruit, —Ph. G. II............alco., dry	oz. .50
" do................aqu., "	oz. .50

☞ **When ordering, specify: "MERCK'S"!**

Extracts,—*continued:*
—[*Fluid Extracts, see pages 61–63!*]—

		Containers incl.
Colocynth — (as above!), —	{ compound,—Ph. G. I…dry	lb. 5.50
	" —Ph. Brit. …… soft	lb. 3.50
	" —U. S. Ph. … powder	lb. 4.00
Columbo (Colombo), see Extract, Calumba.		
Condurango (Cundurango) [Mataperro]:		
bark …………………… alco., dry		oz. 1.00
" " ……………………… " soft		oz. 1.00
Conium, see Extract, Hemlock (*Spotted* H.).		
Convallaria, see Extract, Lily of the Valley..		
Corn-silk (Maize-silk) [Stigmata Maydis], [alco., soft		oz. .50
Coto-bark ……………………… aqu., soft		oz. 1.50
Cotyledon umbilicus, see Extract, Navelwort		
Couch - grass (Quick - grass, Dog - grass; Quickens, Quitch): rhizome;—[Extractum Tritici repentis],—Extractum Graminis, Ph. G. II……………… aqu., soft		lb. .75
Crocus, see Extract, Saffron ……………		
Croton eluteria, see Extract, Cascarilla….		
Cubeb: fruit…… ehtereal,—(Oleoresin of Cubeb)		oz. 1.00
" " —Ph. G. II … alcoholo-ethereal		oz. 1.00
" " —Ph. Austr. ……… alcoholic		oz. 1.00
Cucumber, Wild (Squirting), see Extract, Squirting Cucumber ………………		
Cundurango, see Extract, Condurango ….		
Curcuma, see Extract, Turmeric ……….		
Cusso (Kousso), see Extract, Brayera……		
Cutch, see Extract, Catechu …………		
Cynoglossum, see Extract, Hound's tongue.		
Damiana (Turnera aphrodisiaca): leaves; [alco., soft		oz. .50
Dandelion (Taraxacum), freshly dried root and herb,—Ph. G. II ….. aqu., soft		lb. .75
" fresh root and herb …… " "		lb. 1.00
Datura stramonium, see Extract, Stramonium …………………………………		
Deadly Nightshade, see Extract, Belladonna		
Digitalis: dry leaves ………… aqu., soft		lb. 1.35
" fresh " …… from juice, "		lb. 1.50
" " " —Ph. G. II … alco., "		lb. 3.00
" " " —with powd. Licorice-root,-Ph. G. II,-[50% of soft]….. alco., dry		lb. 3.00
" recently dried leaves …… " soft		lb. 3.50
" dry leaves,—green ……… " "		lb. 2.50
Dogwood-bark, Jamaica, see Extr., Piscidia		
Duboisia: leaves ………… aqu., soft		oz. 5.00
Dulcamara, see Extract, Bittersweet …….		
Ecballium-fruit, and juice) see Ext., Squirt- Elaterium-fruit, and juice) ing Cucumber.		
Elecampane: root, (Alant-root, Inula-root; Radix Helenii) ……… aqu., soft		lb. 1.25
" do.,—Ph. G. II………… alco., "		lb. 3.00
English Walnut, (Juglans regia), see Extract, Walnut ………………………		
Ergot of Rye, (Spurred Rye—Secale cornutum [clavatum]); aqu., soft		lb. 4.00
" " " —Ph. G. II,—(the "*Ergotinum*" of Ph. G. II); hydro-alco., soft, depur. by Alco.		lb. 4.50
Erythræa centaurium, see Extr., Centaury, European ………………………		
Erythroxylon, see Extract, Coca ………		
Eucalyptus: leaves … ethereal, soft,—(Oleoresin of Eucalyptus)		oz. .75
" " ………… aqu., soft		oz. .30
" " ………… alco., dry		oz. .40

☞ When ordering, specify: "**MERCK'S**"!

	Containers incl.		
Extracts,—*continued:* —[*Fluid Extracts,* see pages 61–63!]— Fennel, Water-, see Extr., Phellandrium... Fern, male ⎱ (Aspidium), see Extract, Male Filix mas ⎰ Fern..................... Foxglove (Purple Foxglove), see Extract, Digitalis................................			
Frangula (Alder Buckthorn, European Buckthorn): bark..................aqu., dry	lb. 2.00		
Fucus vesiculosus, see Extr., Bladder-wrack			
Fumaria ⎱ : herb...............aqu., soft Fumitory ⎰	lb. 1.50		
Garcinia, see Extract, Mangosteen.........			
Gelsemium (Yellow [Wild] Jessamine): root; [alco., soft	oz. .50		
" do........................ " dry	oz. .75		
Gentian (Gentiana *lutea* [*rubra; major*]*!*): root,—Ph. Brit...aqu., soft	lb. .75		
" " —Ph. G. II.....cold process, [aqu., soft	lb. .65		
" " ......cold process, " dry	lb. 1.50		
" " ....alco., soft	lb. 1.50		
Gentiana (Erythræa; Chironia) *centaurium,* see Extract, Centaury, *European*........			
Glandulæ rottleræ, see Extract, Kamala...			
Glycyrrhiza, see Extract, Licorice-root.....			
Glycyrrhiza, purified, see Extract, **Licorice**...			
Golden Seal, (Hydrastis): root, [Yellow Root, Orange Root, Indian Turmeric]...hydroalcoholic, dry	oz. .75		
Gramen;—(Extractum Graminis, Ph. G. II), —see Extract, Couch-grass.............			
Granatum, see Extract, Pomegranate.....			
Granatum, Java, see Extr., **Pomegranate,**—Java			
Gratiola (Hedge-hyssop): dry herb; aqu., soft	lb. 1.50		
" fresh herb ....alco., "	lb. 3.00		
" " " -green,-Ph. Neer.; "	oz. .50		
Grindelia: flowering herb........aqu., soft	oz. .50		
Guaiacum-wood (Lignum guajaci; Lignum [*not Arbor!*] vitæ; Lignum sanctum); [aqu., soft	oz. .30		
¨ .... " dry	oz. .40		
' ....alco., soft	lb. 1.50		
" .... " dry	lb. 2.00		
Guarana-paste ....alco., dry	oz. 1.50		
Hæmatoxylon, see Extract, Logwood......			
Hamamelis, see Extract, Witch-hazel......			
Hedge-hyssop, see Extract, Gratiola			
Helenium-root (Inula-root), [*not Sneezewort or Sneezeweed!*], see Extract, Elecampane.			
Hellebore, White, European,—see Extract, Veratrum, White.................			
" Black, (Christmas - rose): root, [Radix melampodii].......alco., soft	lb. 1.75		
" " do............aqu., "	lb. 2.50		
" Green, *European,* (Winter Hellebore), [*not Green Veratrum!*]: root,—Ph. Austr............................soft	lb. 3.00		
Hemlock (*Spotted* [Poison] Hemlock), (Conium]: dry herb.........aqu., soft	lb. 1.00		
" fresh herb..........from juice, "	lb. 1.00		
" " " ....alco., "	lb. 2.50		
" " " —with Dextrin, — [50% of soft]..........alco., dry	lb. 2.50		
" dry " —green......... " soft	lb. 3.50		
Hemlock (Conium): fruit [seed]...alcoholic	oz. .60		
Hemlock, *Water*-, Five-leaved, see Extract, Phellandrium......................			
Hemp (Cannabis), Indian, see Extract, Indian Hemp			

☞ When ordering, specify: "**MERCK'S**"!

Extracts,—*continued:*
—[*Fluid Extracts,* see pages 61–63 !]—

	Containers incl.			
Henbane, see Extract, Hyoscyamus.......				
Hoarhound (Horehound) [Marrubium]: herb.......................aqu., soft	lb. 1.00			
Hound's tongue, (Cynoglossum): root.aqu., [soft	lb. 1.50			
Hydrastis, see Extract, Golden Seal.......				
Hydrocotyle (Water-Pennywort, Indian Pennywort): herb........aqu., soft	oz. 1.00			
" do......................alco., "	oz. 1.00			
" " .... " dry	oz. 1.00			
Hyoscyamus: dry leaves.........aqu., soft	lb. 1.50			
" do. do., — with Dextrin, — [50% of soft]...........aqu., dry	lb. 1.50			
" " " —without admixt., " "	lb. 1.75			
" fresh leaves..........from juice, soft	lb. 1.25			
" " " —Ph. G. II......alco., "	lb. 2.50			
" " " —w. Licor.-root,-Ph. G. II,-[50% of soft]..alco., dry	oz. .35			
" " " —with Milk-sugar,—[50% of soft]........alco., dry	oz. .40			
" recently dried leaves....... " soft	oz. .60			
" dry leaves,—green......... " "	oz. .30			
Hyoscyamus: seed..............alco., dry	oz. 1.25			
Ignatia (St. Ignatius's Bean): seed; alco., dry	oz. .75			
Indian Hemp ⎫ ⎰ herb; ethereal—(Oleoresin of Indian Hemp)	oz. .60			
" " ⎪(Indian " —Ph.G.II..alco., soft	oz. .30			
" " ⎬ " —w. pwd. Licor.-root, —Ph. G. II,—[50% of soft]...alco., dry	oz. .40			
" " ⎪Cannabis) " —w. Milk-sug.,-[50% of soft]...alco., dry	oz. .40			
" " ⎭ " —w. Dextrin,—[33⅓% of soft]...alco., dry	oz. .40			
Indian Pennywort, see Extr., Hydrocotyle				
Indian Tobacco, see Extract, Lobelia......				
Inula-root, see Extract, Elecampane......				
Ipecac (Ipecacuanha): root........aqu., dry	oz. .90			
" do................hydro-alcoh., "	oz. 2.00			
Ipecac: root,—alcoholic,—see Emetine.......				
Iron malate, so-called,—(Extractum ferri pomatum, Ph.G. II),—see Extract, Apple, [ferrated				
Jaborandi (Pilocarpus): leaves....aqu., dry	oz. .50			
Jalap: root (tuber); true........aqu., soft	lb. .75			
" " " .... " dry	lb. 2.00			
Jamaica Dogwood, see Extract, Piscidia ..				
Jessamine, Wild (Yellow), see Extr., Gelsemium				
Juglans regia, see Extract, Walnut........				
Juniper: fresh fruit (berries),—inspissated infusion; — [Succus Juniperi inspissatus]..............................soft	lb. .30			
Kamala (Kameela) [Rottlera tinctoria]: capsule-glands; (Glandulæ rottleræ); [alco., dry	oz. 1.50			
" do...ethereal,—(Oleoresin of Kamala)	oz. 1.50			
Kava-kava (Ava): root......hydro-alcoholic	oz. 1.00			
Kousso (Kooso, Cusso), see Extract, Brayera				
Krameria,—*U. S. Ph.*, and others,—see Extract, Rhatany, etc.....................				
Lactuca virosa, see Extract, Lettuce.......				
Lactucarium; — (Extract from Germanic Lactucarium, ⎱ Purified [from the so-called "Lettuce opium"]),-alco., soft ⎰ Lactucarium...	oz. 1.25			
" .... " dry	oz. 1.25			
Lappa, see Extract, Burdock.............				

☞ When ordering, specify: "**MERCK'S**"!

MERCK'S INDEX. 55

	Containers incl.			
Extracts,—*continued*:				
—[*Fluid Extracts*, see pages 61-63 l]—				
Lettuce ⎤ ⎛ dry leaves .... aqu., soft	lb. 2.25			
" ⎟ ⎜ fresh " ...from juice, "	lb. 2.50			
" ⎟ ⎜ " " -Ph.G. I,-alco., "	lb. 3.00			
" ⎥ ⎜ " " -w.Lic.-r.,-[50%of				
⎟ ⎜ soft]...alco., dry	lb. 4.00			
" ⎦ ⎝ dry " -green ; alco., soft	lb. 4.00			
Levant Wormseed, (Cina; Artemisia maritima): flower-buds, — [Santonica ; Semen - contra] ;				
[ethereal, soft	oz. .40			
" " do................alco., "	oz. .40			
Levisticum, see Extract, Lovage..........				
Licorice (Liquorice),—perfectly clearly soluble, —from the crude extract;—(Purified Extract of Glycyrrhiza)........**soft**	lb. .70			
" from the crude extract...........dry	lb. 1.00			
Licorice-root (Glycyrrhiza); cold proc., soft	lb. 1.50			
" .... " " dry	lb. 2.00			
Licorice-root,–purified,–see Extract, **Licorice**.				
Lignum vitæ (sanctum), [*not Arbor vitæ !*], see Extract, Guaiacum-wood............				
Lily of the Valley, (Convallaria): entire plant; [aqu., dry	lb. 2.00			
" " " do............. " soft	lb. 1.90			
" " " " ....alco., "	lb. 2.50			
Liquorice, and Liquorice-root, see Extr., **Licorice**, and Licorice-root..............				
Lobelia (Indian Tobacco): herb...alco., soft	oz. .50			
Logwood(Hæmatoxylon;Campeachy-wood); [aqu., dry, officinal	lb. 1.50			
" .... " commercial, I	lb. .50			
Lovage (Levisticum): root.......alco., soft	lb. 3.00			
Lupuline (the glandular powder from Hopcones) ....aqu., soft	lb. 1.50			
" ....alco., "	lb. 1.50			
" .... " dry	lb. 1.50			
Madder (Rubia): root............aqu., soft	lb. 2.00			
Maize-silk (Stigmata Maydis), see Extract, Corn-silk				
Male Fern, (Aspidium filix mas): rhizome;— ethereal,—(Oleoresin of Aspidium, *U. S. Ph.*),—[sometimes called "Liquid Extr. of Male Fern," or "Oil of Fern"].....	lb. 2.50			
" " do.; –Ph. G. II......ethereal,— [free fr. Ether	lb. 2.75			
" " " —Ph. Austr........alcoholic	lb. 1.50			
Malt, Barley-,—Ph. G. I & II.........soft	lb. .75			
" " ....dry, powder	lb. 1.25			
" " —lupulated (hopped)......soft	lb. 1.00			
Mandrake (May-apple; Podophyllum): root [rhizome],—*U. S. Ph.*..........alco., soft	lb. 2.50			
Mangosteen (Garcinia): fruit-rind..aqu., dry	oz. .80			
Marigold, Garden, see Extract, Calendula..				
Marrubium, see Extract, Hoarhound......				
Marsh Trefoil, see Extract, Buckbean				
Mary-Thistle (Carduus Mariæ): seed...aqu.	oz. .75			
Mataperro, see Extract, Condurango......				
Matico: leaves..................ethereal, —(Oleoresin of Matico)	oz. .75			
" " .... aqu., soft	oz. .40			
" " .... alco., "	oz. .40			
Matricaria, see Extract, Chamomile, German				
May-apple,—*U. S. Ph.*,—see Extract, Mandrake				
Meadow-saffron, see Extract, Colchicum...				
Melampodii radix, see Extract, Hellebore, Black: root........................				

☞ When ordering, specify : "**MERCK'S**"!

	Containers incl.			
Extracts,—*continued:* —[*Fluid Extracts*, see pages 61–63!]—				
Menyanthes trifoliata, (Marsh Trefoil), see Extract, Buckbean				
Mezereon (Spurge Olive): bark....ethereal,				
—(Oleoresin of Mezereon)	oz. .75			
" do............alco., soft ⎫ (*Mezerein*)	oz. .40			
" "........" dry ⎭	oz. .50			
Milfoil (Millefolium; Achillea), see Extract, Yarrow				
Milkwort, Bitter, European, see Extr., Polygala amara				
Momordica elaterium: fruit, and juice,— see Extr., Squirting Cucumber				
Monesia-bark.............aqu., dry	oz. .40			
Monkshood, see Extract, Aconite				
Mugwort (Artemisia vulgaris): root..alco., [soft	oz. .40			
Myrobalan: fruit...........aqu., dry	oz. .40			
Myrrh..................aqu., dry	lb. 3.00			
" ....aqu., scales	lb. 4.00			
Navelwort (Pennywort) [Cotyledon]: herb; [soft	oz. 1.00			
Nicotiana, see Extract, Tobacco				
Nux vomica, (Semen Strychni), [Poison-nut].........aqu., dry	oz. .20			
" " by Alc. of 0.894,–Ph. G. II,– dry	oz. .30			
" " " " " 0.892,–Ph. Austr.,– soft	oz. .30			
" " " " " 0.879,–Ph. Neerl.,– soft	oz. .30			
" " " " " 0.838,–Ph. Br.'67,– soft	oz. .35			
" " " " " 0.884,– " " new,— [15% Alkaloid],–soft	oz. .35			
" " w.Milk-sug., ⎰ [50% of soft] ⎱ dry	oz. .40			
" " " Dextrin, ⎱ —Ph. Aust. ⎰ "	oz. .40			
Oak-bark................aqu., dry	lb. 2.00			
Opium,—Ph. G. II...........aqu., dry	oz. 1.00			
" ...." soft	oz. .77			
" w.Dextrin,—[50% of soft],— " dry	oz. 1.00			
Orange, Bitter, see Extract, Bitter Orange				
Papaveris capitum, see Extract, Poppy-heads				
Pellitory, German, (Pyrethrum germanicum): root..............alco., soft	oz. .65			
Pennywort (Cotyledon umbilicus), see Extr., Navelwort				
Pennywort, Water-, (Indian Pennywort), see Extr., Hydrocotyle				
Pepper, Black: fruit........alco., soft	oz. 1.50			
Pepper,—Red (Pod, Cayenne); and African [Guinea, Bird],— see Extract, Capsicum annuum; and, fastigiatum				
Phellandrium (Water-Fennel; Five-leaved Water-Hemlock): fruit....ethereal, — (Oleoresin of Phellandrium)	oz. .60			
" do................aqu., soft	oz. .30			
" "................alco., "	oz. .50			
Physostigma (Calabar Bean): seed; alco., dry	oz. 1.50			
" do................." soft	oz. 1.25			
" "..........alcoholo-acetic, "				
Pilocarpus, see Extract, Jaborandi				
Pimpinella-root..........alco., soft	lb. 3.00			
" ....aqu., "	lb. 2.50			
Pine-needles (Leaves of Pinus sylvestris)	lb. .60			
Piscidia (Jamaica Dogwood): bark; alco., dry	oz. 1.00			
Podophyllum,—*U. S. Ph.*,—see Extract, Mandrake				
Poison-nut, see Extract, Nux vomica				
Poison-oak (Rhus toxicodendron): leaves; [alco., soft	oz. .30			
" do...............aqu., "	oz. .25			

☞ **When ordering, specify: "MERCK'S"!**

MERCK'S INDEX. 57

	Containers incl.			
Extracts,—*continued:*				
—[*Fluid Extracts,* see pages 61–63 !]—				
Polygala amara, (European Bitter Polygala; European Bitter Milkwort): entire plant; [aqu., soft	lb. 2.00			
Polygala senega, see Extract, Senega				
Pomegranate (Granatum): root-bark..aqu., [dry	oz. .35			
" do........................alco., soft	oz. .30			
Pomegranate: fresh root-bark,–Java,–alco., soft	oz. 2.00			
Poplar-buds (Gemmæ populi), fresh..aqu., [soft	oz. .50			
" do........................alco., "	oz. .45			
Poppy-capsules (-heads).........aqu., soft	lb. 1.75			
" ....alco., "	lb. 3.00			
Pulsatilla (European Meadow Anemone): dry herb...............aqu., soft	lb. 2.00			
" " " —green........alco., "	lb. 4.50			
" fresh " —Ph. G. I...... " "	lb. 5.00			
Pyrethrum germanicum, see Extract, Pellitory, German				
Quassia-wood (Bitter Wood, Bitter Ash); [aqu., soft	lb. 3.00			
" —Ph. G. II............... " dry	oz. .50			
" ....alco., "	oz. 1.00			
Quebracho blanco: bark:—				
aqueous, dry.........................	oz. 1.00			
alcoholic, "	oz. 1.00			
according to Penzoldt,–liquid;–(*Tincture!*)	lb. 3.00			
" " " –dry.............	oz. 1.25			
Quebracho colorado: wood:—				
aqueous, dry.........................	oz. .30			
" liquid	oz. .25			
Quercus marina, see Extr., Bladder-wrack..				
Quick-grass (Quickens, Quitch) [Triticum repens], see Extract, Couch-grass				
Quillaya (Quillaia saponaria): bark, [Soap-bark].....................aqu., soft	lb. 3.50			
Quince, Bengal, see Extract, Bael, Indian.				
Quinine-plant (Quinine-flower) [Sabbatia Elliottii]: herb..............aqu., soft	oz. .75			
Rhamnus frangula, see Extract, Frangula.				
Rhamnus purshiana: bark, see Extr., Cascara sagrada				
Rhatany (Ratanhia; Krameria): root..cold [process, aqu., dry,—I	lb. 2.75			
" do........cold process, " " -II	lb. 1.50			
" " " " " scales	lb. 2.50			
" " ....alco., dry	lb. 3.00			
" " —*Extractum Krameriæ, U. S. Ph.*; [cold process, aqu.. dry	lb. 1.50			
Rhubarb, Asiatic: root...........aqu., dry	oz. .25			
" " " ....alco.,soft	oz. .25			
" " " -Ph. G. II.. " dry	oz. .40			
Rhubarb, Asiatic, — compound,— Ph. G. II	oz. .35			
Rhus toxicodendron, see Extr., Poison-oak.				
Rottlera (Glandula rottleræ), see Extract, Kamala.........................				
Rubia, see Extract, Madder...............				
Rue (Ruta): leaves ....aqu., soft	lb. 2.25			
" do. ....alco., "	lb. 3.00			
Sabbatia Elliottii, see Extr., Quinine-plant.				
Sabina, see Extract, Savin................				
Saffron (Crocus)................alco., soft	oz. 3.50			
Saffron, Meadow-, see Extract, Colchicum.				
Saint-Ignatius's Bean, see Extract, Ignatia..				
Salix, see Extract, Willow................				
Sanguinaria, see Extract, Bloodroot.......				
Santonica (Flores Cinæ; "Semen Cinæ"), see Extr., Levant Wormseed				

☞ When ordering, specify: "**MERCK'S**"!

Extracts,—*continued:*
—[*Fluid Extracts*, see pages 61–63 1]—
Saponaria officinalis, see Extract, Soapwort

	Containers incl.			
Sarsaparilla..................aqu., soft	lb. 2.25			
" .... " dry	oz. .40			
" ....alco., soft	lb. 3.50			
" .... " dry	oz. .50			
Sassafras-root (Lignum Sassafras); aqu., soft	lb. 3.00			
Savin (Sabina): dried tops........aqu., soft	lb. 1.75			
" do.,—Ph. G. H..hydro-alcoholic, soft	lb. 2.50			
Scilla, see Extract, Squill...............				
Scurvy-grass (Spoonwort) [Cochlearia], fresh herb.................from juice, soft	lb. 2.50			
Sea-wrack (Fucus vesiculosus), see Extract, Bladder-wrack.........................				
Secale cornutum (clavatum), see Extr., Ergot of Rye.................................				
Semen-contra (Santonica), see Extr., Levant Wormseed......................				
Senega: root, (Senega Snakeroot), [Radix Polygalæ senegæ].......aqu., dry,	oz. 1.00			
" do.......................alco., "	oz. .75			
Senna: leaves ....aqu., soft,	lb. 1.75			
" " ....alco., "	lb. 1.75			
Serpentary (Serpentaria): rhizome, [Virginia Snakeroot]....................alco., soft	oz. 1.25			
Shamrock, Water-, see·Extract, Buckbean.				
Simaruba: bark.................aqu., soft	oz. .75			
" " ....alco., "	oz. 1.00			
Snakeroot, Black, (Cimicifuga), see Extract, Black Cohosh				
Snakeroot, Senega, see Extract, Senega....				
Snakeroot, Virginia, see Extract, Serpentary				
Soap-bark, see Extract, Quillaya.........				
Soapwort (Saponaria officinalis): root, [Soap-root]....................aqu., soft	lb. 1.50			
" do......................alco., "	lb. 3.00			
Spanish Flies, see Extract, Cantharides....				
Spoonwort (Cochlearia), see Extr., Scurvy-grass...................................				
Spurge Olive, see Extract, Mezereon......				
Spurred Rye, see Extract, Ergot of Rye....				
Squill (Scilla): dried bulbs.......aqu., soft	lb. 1.00			
" do. do.................. " dry	lb. 1.50			
" " " —Ph. G. II.........alco., soft	lb. 1.50			
Squirting Cucumber, (Wild Cucumber), [Ecballium (Momordica) elaterium]: nearly ripe fruit............. aqu., soft } Elaterium nigrum verum, (True Black Elaterium).	oz. .50			
Squirting Cucumber: fresh juice of the fruit, —Ph. Austr..................alco., soft	oz. 1.00			
N. B.— *Compare, also:* Elaterium (*Elaterium Clutterbuck*).				
Stigmata Maydis, (Maize-silk), see Extract, Corn-silk				
Stramonium (Datura S.): dry leaves..aqu., [soft	lb. 1.35			
" fresh leaves..........from juice, "	lb. 1.75			
" " " ....alco., "	lb. 2.00			
" " " —w. Lic.-root,—[50% of soft],—alco., dry	lb. 2.50			
Stramonium: seed ....alco., dry	oz. 1.25			
Strychnos-seed, see Extract, Nux vomica..				
Succory, see Extract, Chicory, Wild.......				
Sweet Flag, see Extract, Calamus.........				
Sweetwood (Croton eluteria), see Extract, Cascarilla............................				
Taraxacum, see Extract, Dandelion........				
Tetterwort, see Extract, Celandine				
Thistle, Blessed, see Extr., Blessed Thistle.				

☞ **When ordering, specify: "MERCK'S"!**

Extracts,—*continued:*
—[*Fluid Extracts*, see pages 61-63!]—

	Containers incl.		
Thistle, Mary-, see Extr., Mary-Thistle....			
Thornapple, see Extract, Stramonium			
Tobacco (Nicotiana): dry herb....aqu., soft	oz. .35		
" do. do. ....alco., "	oz. .40		
Tormentil: root (rhizome)........aqu., dry	lb. 3.50		
Toxicodendron (Rhus toxicodendron), see Extract, Poison-oak			
Trifolium fibrinum, (Menyanthes trifoliata), see Extract, Buckbean			
Triticum repens, see Extract, Couch-grass..			
Tschuchiakabi (a Japanese Orchidea): fruit			
Turmeric (Curcuma): root [rhiz.]; alco., soft	oz. .50		
Turnera aphrodisiaca, see Extract, Damiana....			
Uva ursi (Uvæ ursi *folia*), see Extract, Bearberry: leaves.......................			
Valerian: root (rhizome) ..ethereal,—[Oleoresin of Valerian]	oz. .75		
" " ....cold process, aqu., soft	lb. 2.00		
" " ....aqu., soft I.	lb. 1.75		
.. " ...." " II.	lb. 1.00		
" " —Ph. G. I......alco., soft	lb. 2.50		
Veratrum, White, (European White Hellebore): root [rhizome] ....alco., soft	oz. .30		
Viburnum (V. prunifolium), see Extract, Black Haw			
Vomic-nut (Semen Strychni), see Extract, Nux vomica..			
Walnut (English Walnut) [Juglans regia]: pericarp ....aqu., soft	lb. .75		
" " ....alco., "	lb. 2.00		
" " —Ph. Ross. ....dry	lb. 2.00		
Walnut,-as above: leaves.......aqu., soft	lb. 1.25		
" " " ....alco., "	lb. 2.00		
Water-Fennel (Five-leaved Water-Hemlock), see Extract, Phellandrium.............			
Water-Pennywort, see Extract, Hydrocotyle			
Water-Shamrock, see Extract, Buckbean...			
Wild Cucumber, see Extract, Squirting Cucumber			
Wild Jessamine, see Extract, Gelsemium ..			
Willow (Salix, divers species): bark; aqu., dry	lb. 1.75		
Witch-hazel (Hamamelis): bark.....hydroalcoholic, dry	oz. .75		
N.B.—*Compare, also:* **Hazeline!**			
Wolfsbane, see Extract, Aconite..........			
Wormseed, Levant-, (Santonica), see Extr., Levant Wormseed			
Wormwood (Absinthium; Artemisia absinthium): herb............aqu., soft	lb. 1.00		
" do,—Ph. G. II...........alco., "	lb. 2.00		
Yarrow (Milfoil, Millefolium; Achillea): flowering herb.........aqu., soft	lb. 1.00		
" do. do...................alco., "	lb. 2.50		
Yellow Jessamine, see Extract, Gelsemium.			

Extracts, Fluid, see *Fluid Extracts*;—pages 61-63.

Extractum Fellis bovini, (*Extract of Ox Gall*), see Gall, Ox-, inspissated, *U. S. Ph.*..

☞ When ordering, specify: "**MERCK'S**"!

	Containers incl.			

	Containers incl.		
Fluid Extracts,—(inserted in alphabetical place of *Extracts, Fluid*):— [Unless otherwise specified, these Extracts are prepared according to the formula of the *United-States Pharmacopœia:* — "Proportion of the crude drug to the extract = 100 grammes : 100 cubic centimetres."] *From:*			
Absinthium (Wormwood): herb .. Artemisia [absinth.	lb. 2.50		
Adonis vernalis, (Bird's Eye; False Hellebore): herb	lb. 3.50		
Anemone, European Meadow-, see Fluid Extract, Pulsatilla			
Arbor vitae, [*not Lignum vitæ!*], see Fluid Extract, Thuja			
Arnica-root............Arnica montana	lb. 2.25		
Aurantii cortex, (Bitter-Orange peel).......	lb. 2.50		
Bela (Indian Bael, Bengal Quince): fruit ..	lb. 2.00		
" do.,—Ph. Brit......................	lb. 1.85		
Belladonna-root........................	lb. 1.75		
Berberis aquifolia, (Holly-leaved Barberry —*not Bearberry!*): root..............	lb. 2.25		
Buchu (Bucco): leaves...Barosma, div. spec.	lb. 2.00		
Bursa pastoris, (Capsella B. p.),-[Sheperd's purse]: fresh herb.—(N. B.—Only preparations from the *fresh herb* possess the remarkable hemostatic virtues of this plant.)	lb. 2.50		
Cahinca-root (Radix *caincæ [cainanœ]*); Chio- [cocca racemosa	lb. 2.50		
Calendula (Garden Marigold): flowers...C. [officinalis	lb. 5.00		
Calumba (Columbo): root ....Cocculus pal- [matus	lb. 1.50		
Cannabis indica, (Indian Hemp): herb.....	lb. 2.25		
Capsella bursa pastoris, see Fl. Extr., Bursa pastoris			
Capsicum (Red Pepper): fruit...C. annuum	lb. 1.75		
Cascara sagrada, (Chittem-bark)....Rham- [nus purshiana	lb. 3.00		
Chamomile - flowers, German, (Matricaria); [Chamomilla vulgaris	lb. 2.00		
Chicory, Wild, (Succory): root...Cichorium [intybus	lb. 1.75		
Cimicifuga (Actæa) [Black Cohosh]: root; [C. racemosa	lb. 1.75		
Cinchona-bark, Gray.....................	lb. 2.25		
" Pale........................	lb. 2.25		
" Succirubra.....................	lb. 2.50		
" Yellow, (True Calisaya-bark — Cortex cinchonæ regiæ);—sp. gr. 1.1......	lb. 3.00		
Coca (Erythroxylon): leaves	lb. 2.00		
Cola-nut (Guru-nut, Caffeine-nut).........	lb. 3.00		
Colchicum (Meadow - saffron): root [bulb]; [C. autumnale	lb. 2.00		
Colchicum: seed............ " "	lb. 2.25		
Colocynth (Bitter Apple): fruit....Cucumis [colocynthis	lb. 4.00		
Condurango (Mataperro): bark ......Gono- [lobus condurango	lb. 2.00		
Convallaria majalis: entire plant	lb. 1.50		
Corn-silk (Maize-silk) [Stigmata Maydis]; [Zea mays	lb. 4.00		
Coto-bark, Para-........................	lb. 3.00		
Cubeb: fruit ....Cubeba officinalis	lb. 4.00		
Damiana: leaves.....Turnera aphrodisiaca	lb. 2.00		
Dulcamara (Bittersweet): young branches; [Solanum dulcamara	lb. 2.00		
Ergot of Corn, (Corn-ergot, Corn-smut), [Ustilago maydis]	lb. 3.00		

☞ When ordering, specify : "MERCK'S" !

	Containers incl.		
Fluid Extracts, — (inserted in alphabetical place of *Extracts, Fluid*), — *continued:* — [Other *Extracts*, see pages 48–59!] —			
Ergot of Rye, (Spurred Rye — Secale cornutum), — *U. S. Ph.*	lb. 1.85		
" " " — Ph. Brit.	lb. 2.00		
Eucalyptus globulus: leaves	lb. 2.25		
Euonymus (*Evonymus*) [Wahoo, Spindle-tree, Burning Bush]: bark...E. atropurpureus	lb. 2.50		
Euphorbia pilulifera: herb	lb. 4.00		
Fabiana (Pichi): branches.....F. imbricata	lb. 5.00		
Francisceа (Manacá): root.......F. uniflora	lb. 4.50		
Fucus vesiculosus, (Bladder-wrack), [Quercus marina]	lb. 1.75		
Gelsemium (Yellow Jessamine): root....G. sempervirens	lb. 1.75		
Gentian-root	lb. 1.75		
Gossypium herbaceum: bark of root, (Cotton-root bark)	lb. 1.50		
Grindelia robusta: flowering herb	lb. 1.75		
Guarana-paste, — fr. seed of Paullinia sorbilis	lb. 5.00		
Hamamelis (Witch-hazel): leaves....H. virginica	lb. 1.50		
Hellebore, Green, *European*, (Winter Hellebore), [*not Veratrum viride!*]: root	lb. 2.50		
Hydrastis (Golden Seal): root..H. canadensis	lb. 1.75		
Hyoscyamus (Henbane): leaves....H. niger	lb. 2.25		
Ipecacuanha-root....Cephaëlis ipecacuanha	lb. 4.50		
Jaborandi (Pilocarpus): leaves	lb. 1.75		
Jacaranda: leaves..J. procera, (Bignonia copaia [caroba])	lb. 3.00		
Jalap-root, true............Ipomœa purga	lb. 3.00		
Kava-kava: root..Macropiper methysticum	lb. 2.00		
Krameria, see Fluid Extract, Rhatany-root.			
Leptandra: rhizome, (Black-root, Culver's root)......................L. virginica	lb. 1.75		
Lippia: herb................L. mexicana	lb. 4.50		
Lobelia (Indian Tobacco): herb...L. inflata	lb. 1.75		
Manacá, see Fluid Extract, Franciscea.....			
Maryland Pink, see Fl. Ext., Spigelia.....			
Mountain-balm (Yerba santa): leaves and tops......Eriodictyon californicum (glutinosum)	lb. 2.50		
Muira puama. — (Said to be *the strongest* aphrodisiac known.)	oz. 1.25		
Nux vomica, (Strychnos-seed)	lb. 2.25		
Pichi, see Fluid Extract, Fabiana			
Pilocarpus, see Fluid Extract, Jaborandi...			
Piscidia (Jamaica Dogwood), bark...P. erythrina	lb. 1.75		
Poppy-capsules (-heads)..Papaver somnifer.	lb. 4.00		
Pulsatilla (European Meadow-anemone): herb...............Anemone pulsatilla	lb. 2.00		
Quebracho blanco, } liquid (& dry), see under **Quebracho colorado,** } Extr. (*not* Fluid Extr.)			
Quercus marina, see Fluid Extr., Fucus vesiculosus			
Quince, Bengal, see Fl. Extr., Bela.......			
Rhatany-root (Krameria).........Krameria triandra, [Ratanhia peruviana]	lb. 1.75		
Rhubarb (Rheum), Asiatic: root...........	lb. 2.25		
Rhus aromatica, (Sweet Sumach): root-bark	lb. 2.00		
Salix nigra, (Black Willow): bark..........	lb. 2.50		
Sarsaparilla, — compound................	lb. 1.50		
Sarsaparilla, — simple....................	lb. 1.50		
Senna-leaves............................	lb. 1.50		
Serpentaria: rhizome, (Virginia Snakeroot).	lb. 3.50		
Shepherd's purse, see Fluid Extr., Bursa pastoris			

☞ When ordering, specify: **"MERCK'S"!**

	Containers incl.			
Fluid Extracts,—(inserted in alphabetical place of *Extracts, Fluid*),—*continued*: —[Other *Extracts*, see pages 48–59!]—				
Spigelia (Maryland Pink): herb and rhizome..........S. marilandica [lonicera] N. B.—*Compare, also:* Spigeline!				
Squill-bulbs...............Scilla maritima	lb. 2.50			
Stigmas of Maize, (Stigmata Maydis), see Fl. Extract, Corn-silk.....................	lb. 1.75			
Stillingia (Silver-leaf): root, [Queen's root]; [S. sylvatica	lb. 2.25			
Stramonium-leaves......................	lb. 2.50			
Taraxacum (Dandelion): root...T. officinale	lb. 1.50			
Thuja (Arbor [*not Lignum!*] vitæ) [White Cedar]: small branches...T. occidentalis	lb. 2.25			
Ustilago (U. maydis), see Ergot of Corn...				
Uva ursi, (Bearberry —*not Barberry!*): dried leaves..........Arctostaphylos officinalis	lb. 1.50			
Valerian-root.....•.......................	lb. 1.50			
Viburnum (Black Haw): bark...V. prunifol.	lb. 1.60			
Yerba santa, see Fluid Extract, Mountain-balm...................................				

Blank Space below for inserting additional FLUID EXTRACTS; *on pages* 59 *and* 60 for EXTRACTS and other "E" articles; *on page* 65 for "F" articles.

☞ When ordering, specify: "MERCK'S"!

	Containers incl.			
Febrile Powder, James's, see Antimonial Powder, *U. S. Ph.*				
Fecula, iodized, see Starch, iodized.				
Fehling's Solution (Test-solution), see under: Titrated Normal Solutions,—(*at End of List !*).				
Fel Bovis (*Tauri*) *inspissatum, U. S. Ph.,* see Gall, Ox-, inspissated				
" " purificatum (*depuratum*) siccum, see Sodium, choleate				
Ferrid-compounds, see Iron, Sesqui-compounds				
Ferro-compounds, see Iron, Mono-compounds.				
Ferrugo, see Iron, oxide, brown, *pure*.				
Ferrum, and compounds, see Iron, etc.				
Fibrin, from blood.	15 gr. .20			
" " plants, (Gluten Fibrin).	15 gr. .25			
Figuier's Gold-salt, see Gold and Sodium, chloride, cryst.				
Filhos's Caustic, see Potassium, hydroxide, with Lime, [4:1], fused.				
Filicin, see Acid, filicic.				
Flavoring Oils, so-called, see Oils, flavoring				
Flores, etc., = Flowers, etc.—(Flores *stibii* = Flowers of Antimony; Flores *stanni* [*Jovis*] = Flowers of Tin;—etc., etc.)				
Flores virides æris, (Crystallized Verdigris), see Copper, acetate, normal, *U. S. Ph*.				
Flowers of Antimony, (Antimonious Oxide, — Tri-oxide; *by dry process*), are *chemically identical* with the Wet-process Tri-oxide,—[*which see* under Antimony, oxide, precipitated].				
" of **Arsenic,** resublimed, see Acid, arsenious, etc.				
" of **Benzoin,** see Acid, benzoic, from Siamese (*etc.*) Benzoin-resin ; sublimed,—*U. S. Ph.* ;—and other grades...				
" of **Sulphur,** see Sulphur, sublimed, *U. S. Ph.*				
" of do., washed, see Sulphur, sublimed, washed, *U. S. Ph.*				
" of **Tin,** see Tin, oxide, white, pure . . .				
" of **Verdigris,** (Crystallized Verdigris), see Copper, acetate, normal, *U. S. Ph.*				
" of **Zinc,** see Zinc, oxide, by dry process				
Fluid Extracts—(are inserted in alphabetical place of: *Extracts, Fluid*)—see *pages 61–63*.				
Fluorescein (Resorcin-phtalein).	oz. 1.50			
Fluorescin (Resorcin-phtalin).	oz. 1.25			
Folia Sennæ sine resina, see Senna-leaves, deresinated,—powdered.				
Form-amide	oz. 1.50			
Fowler's Solution, arsenical, see Solutions: Potassium arsenite, *U. S. Ph.*				
Fraxinin (Sugar of Manna), see Mannit. . . .				
Fruit and Flavoring Ethers :				
No. 1. No. 2. No. 3. No. 4.				
Apple. . . . . . " " " "				
Apricot .. " " " "				
Banana. . . . " — — —				
Cherry. . . . . " " " "				
Currant. . . . " " " "				
Gooseberry. " " " "				
Grape . . . . . — " " "				
Lemon. . . . . — " " "				
Orange. . . . . " " " "				

					Containers incl.			
Fruit and Flavoring Ethers,—*continued:*								
	No. 1.	No. 2.	No. 3.	No. 4.				
Peach......	"	"	"	"				
Pear.......	"	"	"	"				
Pineapple..	"	"	"	"				
Quince.....	"	—	—	—				
Radish.....	—	"	—	"				
Raspberry..	"	"	"	"				
Strawberry.	"	"	"	"				
Rum..................................								
Whiskey...............................								
Fruit-sugar I, (Levulose, Lævulose)......					oz. 1.00			
" commercial, (Inverted Sugar),—consisting of Fruit-sugar and Grape-sugar..					lb. .40			
Fuchsine, see under Aniline and Phenol Dyes: Red........								
Furfural (Furfur-aldehyd; Furfurole), chem. pure,—boil.-pt. 160–162° C [320–323.6 F]...					oz. 2.00			
Furfurine................................					15 gr. .50			
" nitrate.............................					15 gr. .50			
Fusel-oil, so-called, see Alcohol, amylic, primary....								
Fusible Metal, see Metal, fusible........								

☞ When ordering, specify: "**MERCK'S**"!

	Containers incl.
Gall, Ox-, (Fel Tauri [Bovis]), purified, dry, see Sodium, choleate..........	
" " inspissated, (Extractum Fellis bovini — Extract of Ox Gall), conforming to U. S. Ph. and Ph. G. I	lb. 1.25
Gallein (Pyro-gallol-phtalein).............	15 gr. .75
Gallium, metallic......................	1½ gr. vial 25.00
Gelatin (*Pure Glutin*), sterilized, for bacteriological purposes......................	oz. 3.50
Gelatin from **Cartilage**, see Chondrin....	
Gelatin, medicated, — in sheets, — see under Atropine and Physostigmine..	
" Discs, medicated, see under Atropine; Cocaine; Duboisine; Physostigmine.	
Gelsemin..............................	oz. 2.50
Gelseminine, — according to Sonnenschein....	15 gr. 2.50
" hydrobromate, amorphous..............	15 gr. 2.50
" hydrochlorate, amorphous.............	15 gr. 2.50
" " cryst., white	15 gr. 3.50
" nitrate, amorphous	15 gr. 2.50
" sulphate, amorphous...................	15 gr. 2.50
Gentian Violet, see under Aniline and Phenol Dyes: Violet.....................	
Gentianin, — extract-form, — (Crude *Gentiopicrin*)..................................	oz. 1.00
Gentisin (Gentianic [Gentisic] Acid).......	15 gr. 2.50
Glass, liquid and soluble, (*Water-Glass*), see Potassium, silicate, etc.; — and, Sodium, silicate, U. S. Ph.; etc., etc...............	
Glass, antimonial, see Antimony, sulphide, vitreous, — so-called	
" Arsenic-, see Acid, arsenious, — *lumps*	
" Borax-, see Sodium, bi-borate, fused	
Glass-etching Ink, see Diamond Ink, so-called...	
Glass-wool, for filters.....................	oz. 1.50
Glauber's Salt, see Sodium, sulphate, (*etc.*).	
Globulin (Crystallin).................?....	15 gr. .50
Globulin, para-, (para-Globulin), pure....	
Glucinum, see Beryllium.................	
Glucose, see **Grape-sugar, chem. pure**; etc......	
Gluten, vegetable......................	oz. 2.50
Glutin, animal, — for use in the arts.......	lb. 2.00
" do., *pure*, — sterilized, — see Gelatin, etc.	
Glycerin (Glycerol), crude, — [26° Baumé], sp. gr. 1.21	
" for gas-meters, — [18° Bé]	
" refined, I, [24° Bé], sp. gr. 1.19	lb. .42
" " " [28° "], " 1.23	lb. .45
" " " [30° "], " 1.25	lb. .48
" " pure, [24° "], " 1.19, redistil.	lb. .45
" " " [28° "], " 1.23,	lb. .48
" " " [30° "], " 1.25, " — U. S. Ph...............	lb. .50
" Price's Patent, — in original 1-lb. bottles.	lb. .75
Glycerin Salicylate, see Ether, glycer.-salic.	
Glycerin, sulphurous, (Solution of Sulphur Di-oxide in Glycerin), [Glycerolate (Glycerite) of Sulphurous Acid]..........	lb. 1.50
Glycerolate of Aluminium acetate, see Aluminium, aceto-glycerolate...........	
N.B.— *Other Glycerolates* — (the class of *Glycerita* or "Glycerites" of the *U. S. Ph.;* and similar preparations, also called Glycerols or Glycerines, — miscalled "Glycerides"; — *all* being simple solutions of active substances in Glycerin, — *not* [as the *real* Glycerides] chemical compounds with Glycerin!): — *see likewise* under the names of their active substances.	

☞ **When ordering, specify: "MERCK'S"!**

	Containers incl.			
Glycium, see Beryllium				
Glycocoll (Glycine, Glycocine; Amido-acetic or Amido-glycollic Acid)................	15 gr. 1.00			
Glycogen (so-called "Animal Amylum"), chem. pure...........................	15 gr. 1.00			
Glycos-amine, hydrochlorate, cryst......	15 gr. 1.50			
Glycyrrhizin, ammoniated, — *U. S. Ph.*, — (*Pharmacopeial* Glycyrrhizate of Ammonium), — soluble............................	oz. .35			
Gold (Aurum), double salts of, see "Gold and —" (below!)...................				
" metallic, powder....................	15 gr. 1.75			
" " precipitated, pure,-amorphous;- soft, lustreless, brown powder.	15 gr. 1.75			
" " do., do., — in fine scales; — with metallic lustre...............				
" bromide	15 gr. 1.50			
" chloride, cryst., yellow..............	15 gr. .75			
" " " brown.............	15 gr. .75			
" " —solution [1:9].............	15 gr. .75			
" cyanide	15 gr. 2.50			
" iodide	15 gr. 2.00			
" oxide	15 gr. 1.50			
Gold and Cadmium, chloride	15 gr. 1.00			
" and **Calcium,** "	15 gr. 1.00			
" and **Potassium,** "	15 gr. 1.00			
" " " cyanide................	15 gr. 1.00			
" and **Sodium,** chloride,—for photography.................	15 gr. .45			
" " do., do.,-*U. S. Ph.*,-[32.4% Gold].	15 gr. .55			
" " " " -Ph. G. II,-[30.3% "].	15 gr. .50			
" " " " cryst., (Figuier's *Gold-salt*)	15 gr. 1.00			
Gold, Alumina Purple of				
" **Figuier's Salt** of, see Gold and Sodium, chloride, cryst.				
" **Tin-precipitate** (*Stannic* precipitate) of,—[Cassius's Purple]............	15 gr. .50			
Goulard's Extract, so-called, (Vinegar of Lead), see Solutions: Lead acetate, basic, *U. S. Ph.*				
Granatin (Sugar of Manna), see Mannit....				
Granella aerophora, see Iron, citrate, effervescent: white or yellow...........				
" do., cum **Magnesia citrica,** see Magnesium, citrate, effervescent, granulated, *U. S. Ph.*				
Grape-sugar (Dextrose, Dextro-glucose, Glucose; Starch-sugar), **chem. pure, anhydrous**..... N. B.—*In contradistinction to other, so-called "chemically pure" brands, which contain as high as 30% of Water,* MY GRAPE-SUGAR, *as above,* IS ABSOLUTELY PURE AND DRY!	lb. 2.00			
do., commercial.......................	lb. .10			
Graphite (Mineral Carbon; Plumbago), purified,—Ph. Bor....................	lb. .75			
" Ceylon............................	lb. .35			
" " finely pulverized, (so-called "*alcoholized*")	lb. .40			
Gregory's Salt, (Hydrochlorate of Morphine and Codeine), see Salt, Gregory's				
Guaiacol (Guajacol), **ch. pure, (absolute),—for medicinal use**;—[Mono-methyl-catechol] .	oz. 1.00			
" commercial.....................	oz. .40			
Guanidine, carbonate, cryst.	15 gr. .25			
Guanine (Guanin)......................	15 gr. 2.00			
" hydrochlorate	15 gr. 1.50			
Guaranine	15 gr. .65			
Gun-cotton, soluble, see Collodion Cotton				
Gutta Percha, purified, white,—in sticks..	oz. .75			

☞ **When ordering, specify: "MERCK'S"!**

	Containers incl.			
Hæmoglobin, Hæmatin, Hæmatoxylin, etc.; see Hemoglobin, Hematin, Hematoxylin, etc.				
Hartshorn, so-called "**Spirit**" of, see Spirit, –so-called,–of Hartshorn........				
Hazeline,—from Witch-hazel (Hamamelis virginica)................................	lb. 2.50			
N.B.—*See, also:*—Extracts: Witch-hazel; —and, Fluid Extracts: Hamamelis.				
Heavy Spar (Barytes), *artificial,* see Barium, sulphate, precipitated, pure.............				
Helenin, cryst., white.—(The **solid** Alant-, or Elecampane-, or Inula-camphor.)—[*Not* to be confounded with *Inulin,*—which *see also!*] N. B.—*Compare, also:* Alantol,—the *liquid* Alant-, or Elecampane-, or Inula-camphor.	15 gr. .50			
Helianthine, see under Aniline and Phenol Dyes: Orange...........................				
Helicin,—from Salicin..................	15 gr. .35			
Helicina, - from snails (Helix pomatia); —[Saccharated Snail-juice]...............	lb. 2.00			
Heliotropin, see **Piperonal,** for perfumery......	15 gr. .35			
Helleborein.—(A *newly discovered use* of this Glucoside is that of a local anesthetic for Ophthalmology. Its anesthesia is reported as considerably *exceeding that of Cocaine* in duration.)	15 gr. 1.00			
Helleborin...............................				
Hematein.—Derivative from Hematoxylin.	15 gr. .50			
Hematin (Hematosin).—Fractional derivative from Hemoglobin	15 gr. 3.00			
Hematoxylin. — The coloring matter of Logwood...............................	⅛ oz. vls.oz. 3.50			
Hemoglobin (Hemato-globulin, Hematocrystallin). — The colored substance of blood	15 gr. .40			

☞ When ordering, specify: "**MERCK'S**"!

		Containers incl.		
Hepar Antimonii (*Stibii*), [Liver of Antimony], see Potassa, antimoniosulphurated, *crude*				
" " **calcareum,** (Calcic Liver of Antimony), see Lime, antimoniosulphurated				
" **Calcis,** (Liver of Lime), see Lime, sulphurated, *U. S. Ph.*				
" **Sulphuris,** (Liver of Sulphur; *Potassic* Liver of Sulphur). see Potassa, sulphurated, *U. S. Ph.;* etc...				
" " **calcareum,** [Calcic Liver of Sulphur], see Lime, sulphurated, *U. S. Ph.*				
" " " **stibiatum,** [Antimonic Liver of Lime; Stibiated Calcic Liver of Sulphur], see Lime, antimonio-sulphurated				
" " **natricum,** (Sodic Liver of Sulphur), see Soda, sulphurated, etc ...., ...				
Hesperetin. — Fractional derivative from Hesperidin	15 gr.	1.50		
Hesperidin. —Glucoside from Oranges	15 gr.	.50		
Hom-atropine Merck - Ladenburg, (Oxy-toluol-tropine):				
pure, cryst.	15 gr.	7.00		
hydrobromate, cryst.	15 gr.	4.50		
hydrochlorate, cryst. ..... *All labels must bear Dr. Ladenburg's (the originator's) signature.*	15 gr.	6.50		
salicylate	15 gr.	6.50		
sulphate, cryst.	15 gr.	6.25		
Hydrargyrum, and compounds, see Mercury, etc				
Hydrastine Merck:				
chem. pure, cryst	15 gr.	.50		
pure, amorphous, powder	15 gr.	.25		
citrate				
hydrochlorate, chem. pure	15 gr.	.50		
nitrate, cryst.,—easily soluble	15 gr.	.60		
phosphate, chem. pure	15 gr.	.60		
sulphate, chem. pure	15 gr.	.50		
tartrate, chem. pure	15 gr.	.50		
Hydro-Berberine, see Berberine, Hydro-				
Hydro-chinone (-*kinone*), see Hydro-quinone				
Hydro-Cotoin, see Cotoin, Hydro-				
Hydrogen Per-oxide (Di-oxide), [Oxygen Hydrate; sometimes called "Oxygenated Water"], medicinal, —aqueous solution [10 volumes of "Active Oxygen"]	lb.	.55		
do. do., commercial, —aqueous solution [10 volumes of "Active Oxygen"]	lb.	.50		
Hydro-quinone — (Hydro-chinone [-kinone])— [Quinol] — (para-Di-oxy-benzene) — [Quinone Hydride]	oz.	.85		
Hydrothion - ammonium, solution, see Solutions: Ammonium sulphide, —hydrosulphuretted				
Hydroxyl-amine, hydrochlorate.	oz.	1.00		
Hyoscine Merck-Ladenburg, —true:				
hydrobromate, cryst .... *All labels must bear Dr. Ladenburg's (the originator's) signature.*	15 gr.	10.00		
hydrochlorate, cryst.	15 gr.	10.50		
hydro-iodate (hydriodate), cryst.	15 gr.	10.00		
sulphate, cryst.				
Hyoscyamine Merck,—true; —from Hyoscyamus niger:				
chem. pure, cryst., white, very light powder,— *U. S. Ph.*	15 gr.	5.00		
pure, not colorless, amorphous	15 gr.	1.75		
hydrobromate, pure, amorphous	15 gr.	1.75		

☞ When ordering, specify: "**MERCK'S**"!

	Containers incl.			
Hyoscyamine Merck. — true; — from **Hyoscyamus niger**;—*continued*:				
hydrochlorate, pure, amorphous............	15 gr. 2.00			
hydro-iodate, (hydriodate), pure, cryst.,—melt.-pt. 154° C [309.2° F].—(The *crystalline* form is *new!*)—[A mydriatic,—more easily *soluble* than the Atropine salt.]....	15 gr. 3.00			
sulphate, pure, amorphous.................	15 gr. 2.00			
" chem. pure. cryst.................	15 gr. 5.00			
Hyoscyamine, derived,—*from Atropine by conversion;* not from Hyoscyamus:				
pure, cryst............................				
hydrobromate, pure, cryst...............				
hydrochlorate, " "				
sulphate..... " "				
Hyper-chlor-acetyl, see **Mono-chlor-ethylene Dichloride**....................................				
Hypnone (Aceto-phenone) [Phenyl-methyl-ketone (-acetone)]	oz. 1.50			
Hypo-quebrachine, see under **Quebracho Alkaloids**				
Hypo-xanthine, see **Sarcine**				

☞ When ordering, specify: **"MERCK'S"!**

	Containers incl.		
Ichthyol preparations:			
Ichthyol-sulphonic (Sulpho-ichthyolic) Acid....	oz. .50		
Ichthyol-sulphonate (Sulpho-ichthyolate) of Ammonium,—[Ichthyol]	oz. .45		
" of Sodium, ...	oz. .50		
" of Lithium........................	oz. .60		
" of Zinc	oz. .50		
Ichthyol Solution, alcoholo-ethereal,—10%	doz. 9.00		
" " " —30%	doz. 12.00		
Ichthyol Plaster, in envelopes............			
(N.B.—*Other Ichthyol preparations,*—such as: Capsules, Pills, Soap, Wadding, etc.,—are furnished by Drug Houses.)			
Ilicin.....................................	15 gr. .50		
Imperatorin, see Peucedanin.............			
Indicator Solutions, (Test-solutions), see at End of List.			
Indigo Blue, see Indigotin................			
Indigo Carmine, best quality,—paste.....	lb. 2.00		
Indigo Sulphate, ("*Soluble Indigo*"), solution, see Tinctures: Indigo...............			
Indigotin (Indigo Blue), pure, cryst.	⅛ oz.vls.oz. 7.00		
Indium, metallic.........................	15 gr. 9.00		
" chloride........................	15 gr. 8.00		
" oxide	15 gr. 9.00		
" sulphate	15 gr. 8.00		
Indole....................................			
Induline, see und. Aniline and Phenol Dyes			
Infernal Stone, see Silver, nitrate, cryst.; and, molded;—*U. S. Ph.;* and, grey........			
Inosit (Meat-sugar)......................	15 gr. 2.75		
Inula-camphor, solid, see Helenin............			
" liquid, see Alantol..................			
Inulin (Alantin, Dahlin; Alant-starch),—according to Dragendorff			
" white			
Inverted Sugar, see Fruit-sugar, commerc'l			
Invertin (Zymase).—The sugar - inverting constituent of yeast.....................	15 gr. 2.00		
Iodine (Iodum), English	lb. 4.10		
" re-sublimed,—*U. S. Ph.* and Ph. G. II.	lb. 4.10		
" chem. pure........................			
" albuminated, (Iodized Albumin).......	oz. 1.00		
" bromide, liquid, (penta-bromide), ["Iodide of Bromine," so-called]			
" chloride (mono-chloride).............	oz. .80		
" tri-chloride.— (Highly efficient antiseptic and disinfectant.)	oz. 1.00		
Iodized Starch, soluble, see Starch, iodized			
Iodo-amyl, see Amyl, iodide			
Iodo-ethyl (Iodide of Ethyl, Mono-iod-ethane), see Ether, hydro-iodic................			
Iodo-methyl, see Methyl, iodide..........			
Iodoform, cryst.,—*U. S. Ph.* and Ph. G. II....	lb. 7.00		
" powder...........................	lb. 7.00		
" " medium grain,-*non-conglutinating*	lb. 7.00		
" " —*so-called* "deodorized" (aromatized).-[For wholly odorless Iodoform, see Iodoform, bituminized.]	oz. .65		
" precipitated	lb. 7.00		
" pencils,—[50% Iodoform]	lb. 7.50		
Iodoform, bituminized (*wholly odorless*).—Translucent scales, easily pulverizable,—totally devoid of the Iodoform odor!	oz. .65		
Iodole (Tetr-iod-pyrrole=C_4I_4NH;—*not*—[as stated in some books:]—"*Tetr-iodide of Pyrrole*" = "$C_4H_8N.I_4$"!).— Contains nearly 89% of Iodine.—[Inodorous, insipid, and non-toxic succedaneum for Iodoform.]....	oz. 1.25		

☞ **When ordering, specify: "MERCK'S"!**

	Containers incl.			
Iodum, and compounds, see Iodine, etc.				
Iridin Merck, pure	oz. 2.00			
Iridium, metallic	15 gr. 2.00			
" " rods	15 gr. 2.00			
" " powder	15 gr. 2.25			
" bromide	15 gr. .50			
" chloride, tri- (sesqui-)	15 gr. 1.00			
" oxide, sesqui-	15 gr. .65			
Iridium and Sodium, chloride, cryst......	15 gr. .75			
Iridium-Osmium alloy, (*Irid-osmium;* Osmiridium), see Osmium-Iridium............				
Iron, **Ferrid**- double salts of, see under Iron, Sesqui-compounds—(below!)				
" **Ferro**- double salts of, see under Iron, Mono-compounds—(below!)				
Iron (Ferrum), metallic, wire,—*U. S. Ph*....	lb. .35			
" do., finely powdered, (so-called "alcoholized"),—Ph.G.II,—(Limatura Martis alcoholisata; Pulvis Ferri alcoholisatus).........................	lb. .35			
" " filings, coarse powder.............	lb. .35			
" " reduced (by Hydrogen), — so-called "Quevenne's Iron,"—[60–65% Iron].............	lb. .73			
" " " —*U. S. Ph*.,—[80% Iron] ...				
" " " chem. pure, - [92–94% Iron].	lb. 2.00			
" " " black,—[50% Iron]	lb. .70			
" acetate, Ferric......................	oz. .25			
" " " in scales.................	oz. .40			
" " " solution, see under Solutions				
" albuminate, (*Iron-Albumin*), in scales, —[5% of Per-oxide—Fe$_2$O$_3$]...	oz. .30			
" " peptonized.....................	oz. .50			
" " saccharated	oz. .40			
N.B.—*Compare, also:* Iron, lactate........ ⎫ albu- " phosphate ..... ⎬ minat- " pyro-phosphate, ⎭ ed.				
" ammoniated, so - called, — (Ammonio - chloride of Iron),—see Ammonium, chloride, with Ferric Chloride.......				
" ammonio-citrate, brown—(*U. S. Ph*.)— or green, see Iron, Sesqui-compounds: Ammonio-Ferric citrate, etc.; etc. ...				
" anisate	oz. 2.50			
" arseniate (arsenate)	oz. .25			
" " —Ph. Brit. new	oz. .25			
" " *and* citrate, ammoniated, [Ammonio-Ferric arsenicico-citrate],— [2% of Arsenicic Acid]	oz. .35			
" arsenite............................	oz. .30			
" benzoate,—[about 25% of Per-oxide]	oz. .50			
" boro-citrate	oz. .50			
" bromide, Ferrous, pure	oz. .22			
" " do., com'l,—[abt. 65-68% Brom.]	lb. 1.00			
" " Ferric, see Iron, tri-bromide				
" bromo-iodide......................	oz. .90			
" by Hydrogen, (reduced),—*U. S. Ph.* and other grades,—see Iron, metallic, reduced, *etc.;* etc......................				
" camphorate	oz. 1.50			
" carbonate, Ferrous, saccharated,—*U. S. Ph*. and Ph. G. I, — [at least 15% of Ferrous carbonate].........	lb. .50			
" " do., do.,—Ph. G. H,—[10% Iron].	lb. .60			
" " green (hydrated)...............	lb. 1.25			
" " sub-, —so-called,—*U. S. Ph.* 1870, —(Aperient Crocus of Iron), see Iron, oxide, brown, (etc.)......				

☞ When ordering, specify: "MERCK'S"!

	Containers incl.			
Iron, chloride, proto- (Ferrous), [Ferrous muriate; di-chloride]	lb. .60			
" " sesqui- (tri-) [Ferric], normal,—cryst., dry; and *U. S. Ph.*; and sublimed, anhydrous;—see Iron, tri-chloride, etc.; *etc.*; etc.				
" " Ferric, basic, (Ferric oxy-chloride), —so-called, - liquid;—see Solutions: Iron oxy-chloride				
" " do., do., dialyzed, see Iron, dialyzed: liquid; and, in scales				
" chromate, liquid	oz. .25			
" citrate,—*U. S. Ph.*,—(Ferric citrate), pure, brown, in scales	lb. 1.00			
" " effervescent, white } granulous powder,—	lb. .95			
" " " yellow } (Granella aerophora)	lb. .90			
" " soluble, so-called, see Iron, Sesqui-compounds: Ammonio-Ferric citrate, in scales: brown—*U. S. Ph.*; and, green				
" " *and* arseniate, ammoniated, see Iron, arseniate *and* citrate, ammoniated				
" citrico-lactate, see Iron, lacto-citrate				
" cyanide, blue, — so - called;—*insoluble;* (Ferro-cyanide of Iron; Ordinary Prussian Blue)	lb. 1.25			
" " blue,—so-called;—*soluble;* (Potassium Ferri-ferro-cyanide; Soluble Prussian Blue)	lb. 1.75			
" dialyzed, liquid, (Ferrum oxydatum dialysatum liquidum,— Ph. G. I),—[Liquid Dialyzed "Basic Ferric Chloride"; Liquid Dialyzed "Ferric Oxychloride",—so-called;—Liquor ferri dialysatus];—[3.5% Iron, = 5% Peroxide]	lb. .35			
" do., in scales	oz. .30			
" ferro-cyanide, (Prussian Blue, ordinary), see Iron, cyanide, blue,—so-called,—*insoluble*				
" granulated sulphate, see Iron, sulphate, Ferrous, pure, precipitated by Alcohol, *U. S. Ph.*				
" hydrate, Ferric, dry } see Iron, oxide,				
" hydrated oxide, Ferric, dry } brown, pure.				
" Hydrogen-reduced,-*U.S.Ph.* and others, -see Iron, metallic, reduced, *etc.;* etc.				
" hypo-phosphite,—*U. S. Ph.*	oz. .25			
" iodate, Ferric	oz. .75			
" iodide, cryst.	oz. .40			
" " insipid	oz. .38			
" " Ferrous, saccharated,—*U. S. Ph.*	oz. .35			
" lactate, pure, cryst., in crusts,-*U. S. Ph.* and Ph. G. II	oz. .18			
" " pure, powder,—Ph. G. II	oz. .15			
" " powder	oz. .12			
" " albuminated	oz. .60			
" lacto-citrate (citrico-lactate)	oz. .35			
" lacto-phosphate (phospho-lactate)	oz. .40			
" malate, in scales	oz. 1.10			
" " crude, see Extracts: Apple, ferrat.				
" metallic, (etc.), see at *top* of "Iron" list				
" oleate	oz. .25			
" oxalate,—*U. S. Ph.*,—Ferrous	oz. .25			
" " Ferric, in scales	oz. .30			
" oxide, black, (Magnetic oxide, Ferrosoferric oxide; Iron Ethiops), —by wet process,—pure	lb. 1.00			
" " " —by dry process	lb. .85			

☞ When ordering, specify : "MERCK'S"!

		Containers incl.		
Iron, oxide, brown, (so-called "sub-carbonate"), [*Aperient* Crocus (Saffron) of Iron],—*Ferri subcarbonas, U. S. Ph.* 1870		lb. .50		
" " " pure, (Dry Hydrated Per-oxide [Sesqui-oxide, Tri-oxide, Red oxide] of Iron; Dry Hydrated Ferric oxide; Dry Ferric Hydrate), —[Ferrugo, Rubigo]		lb. .75		
" oxide, red, (Ferric oxide; Per-oxide, or Tri- [Ter-] oxide, or Sesqui-oxide of Iron), *anhydrous*, — [*Astringent* Crocus (Saffron) of Iron],— (Pure Colcothar, Pure Caput mortuum)		lb. .70		
" " " *do.*,—from Oxalate of Iron.		lb. 2.50		
" " " *hydrated*, dry, see Iron, oxide, brown, *pure*				
" " " peptonated; also, glycerinated solution of *same;*—see Iron, peptonized; etc.— *Same,* dialyzed, see Solutions: Iron, peptonized, dialyzed				
" " " saccharated, soluble, — Ph. G. H; — (so-called "Saccharated Iron" or "Soluble Iron"; Iron Saccharate),—[Ferruginated Sugar; Iron-Sugar];—[3% Iron,=4.285% Per-oxide] N. B.— *See, also: Syrup of* Saccharate of Iron.		lb. .70		
" oxide, dialyzed, (Dialyzed so-called "Ferric Oxy-chloride" or "Basic Ferric Chloride"):- liquid, Ph. G. I,-or, in scales;-see Iron, dialyzed, etc.; etc.				
" oxy-chloride, Ferric, (Basic Ferric Chloride),—so-called ; — solution of,—see under Solutions				
" do., dialyzed, see Iron, dialyzed: liquid; and, in scales				
" peptonized, (Peptonated Ferric Oxide), —clearly soluble in Water,— [2% or 5% Per-oxide]		oz. .35		
" " solution, glycerinated, — for *subcutaneous* injections, — [3 mg Fe₂O₃ and 25 mg Peptone per syringeful]		lb. 1.25		
" " *dialyzed*, liquid,—for *internal* use; —see under Solutions				
" " albuminated, see Iron, albuminate, peptonized				
" " saccharated		oz. .35		
" per-chloride, see Iron, tri-chloride				
" per-oxide, see Iron, oxide, red				
" phosphate,— *so-called by U. S. Ph.*,—see Iron, phosphate, *with Sodium Citrate*..				
" phosphate, *true,* Ferric		lb. 1.00		
" " " Ferrous		lb. .95		
" " albuminated		oz. .35		
" " with Ammonium Citrate, in scales		lb. 1.50		
" " Ferric, *with Sodium Citrate,* in scales,-*Ferri phosphas,* so called by *U. S. Ph.*		lb. 2.00		
" phosphide (phosphuret). — [An indefinite composition of several Iron phosphides.]		oz. 1.00		

☞ When ordering, specify: "MERCK'S"!

	Containers incl.			
Iron, phospho-lactate, see Iron, lacto-phosph.				
" picrate (picro-nitrate)................	oz. .60			
" precipitated sulphate, see Iron, sulphate, Ferrous, pure, precipitated by Alcohol, *U. S. Ph*..............				
" pyro-phosphate,'— *so-called by U. S. Ph*., —see Iron, pyro-phosphate, *with Sodium Citrate*.....................				
" pyro-phosphate, *true*...............	lb. 1.00			
" " albuminated	oz. .65			
" " with Ammonium Citrate, in scales	oz. .30			
" " " Potassium "	oz. .30			
" " " Magnesium " in scales	oz. .35			
" " Ferric, *with Sodium Citrate*, in scales,—*Ferri pyrophosphas*, so called by *U. S. Ph*............	oz. .30			
" reduced (by Hydrogen),—*U. S. Ph*. and other grades,—see Iron, metallic, reduced, *etc.*; etc...............				
" saccharate, ("*Saccharated* Iron" cr "*Soluble* Iron," so-called), see Iron, oxide, red, saccharated.............				
N. B.—*Compare, also:*				
Iron, albuminate .... ⎫				
" carbonate-(*U.S.Ph*.; etc.)- ⎪				
" iodide—(*U.S.Ph*.)—...... ⎬ *saccharated.*				
" peptonized .... ⎪				
" sulphate, Ferrous........ ⎪				
" Mono-compounds: Mangano-Ferrous carbonate .. ⎭				
" salicylate....................	oz. .35			
" santoninate (*not* santonate!),—easily soluble in Alcohol; hardly so in Water	oz. 2.00			
" sesqui-bromide, see Iron, tri-bromide..				
" sesqui-chloride, see Iron, tri-chloride..				
" stearate.........................	oz. .35			
" sub-carbonate, so-called,—*U. S. Ph*. 1870,—(Aperient Crocus of Iron), see Iron, oxide, brown, (etc.)..........				
" sub-sulphate, (Basic Ferric Sulphate), [Monsel's Salt], pure............	lb. .60			
N.B.—*Solution* of do., (*U. S. Ph*.),—[Monsel's Sol.],—see under Sols.				
" succinate........................	oz. .60			
" sulphate, Ferric, normal, (Per-[Sesqui-] sulphate); [Ter-sulphate]......	lb. .40			
" " do., basic, (Monsel's Salt), see Iron, sub-sulphate				
" " Ferrous, pure, (Pure Iron Vitriol; Pure Green Vitriol), cryst., —*U. S. Ph*.............	lb. .25			
" " " pure, (do.; do.), small cryst., —Ph. Neerl............	lb. .30			
" " " pure, precipitated by Alcohol,—Ph. G. II,—("Precipitated Iron," "Granulated Iron,"—so-called),—*Ferri sulphas præcipitatus, U. S. Ph*.................	lb. .30			
" " " pure, calcined (exsiccated, dried),—*Ferri sulphas exsiccatus, U. S. Ph*........	lb. .40			
" " " crude, cryst.,(Crude Iron Vitriol; Crude Green Vitriol)	lb. .20			
" " " saccharated, cryst.	lb. .75			
" sulphide (sulphuret)................	lb. .25			
" " in sticks	lb. .35			
" sulpho-carbolate (phenol-sulphonate, sulpho-phenate)	oz. .20			
" tannate.........................	oz. .25			

☞ When ordering, specify: "**MERCK'S**"!

	Containers incl.			
Iron, tartarated (*tartarized*), see Iron, *Sesqui*-compounds: Potassio-*Ferric* tartrate, *U. S. Ph.*—[*Do not confound* with Iron, tartrate,—(below)!] .... N. B.—*Compare, also:*—Iron, *Mono*-compounds: Potassio-*Ferrous* tartrate,–(*Ferrated Tartar*; Iron-Tartar);—etc.; etc.				
" tartrate, Ferric, in scales ⎫ —[*Do not con*-	oz.	.35		
" " Ferrous .... ⎭ *found* with Iron, *tartarated*,—(above);—nor with Iron-*Tartar*, — (referred-to under same)!]	oz.	.35		
" tri-bromide (sesqui-bromide), [Ferric Bromide], liquid,—sp. gr. 1.400	oz.	.40		
" tri-chloride (sesqui-chloride; per-chloride), [Normal Ferric Chloride], cryst., dry	lb.	.60		
" " cryst.,—*U. S. Ph.* and Ph. G. II,— free from Nitric Acid	lb.	.60		
" " sublimed, anhydrous	oz.	.40		
" " with Ammonium Chloride,—(so-called "Ammoniated Iron "),— see Ammonium, chloride, with Ferric Chloride				
" tri-oxide (ter-oxide), see Iron, oxide, red				
" valerianate, — *U. S. Ph.*	oz.	.35		
Iron,—**albuminated** Oxide or Salts of,—see under Iron : albuminate, etc., etc.; lactate; phosphate; pyro-phosphate .				
" **granulated** .. ⎫ —so-called,—see Iron, sulphate, Ferrous, pure, precipitated by Alcohol. *U. S. Ph.*				
" **precipitated,** ⎭				
" **Quevenne's**, so-called, see Iron, metallic, reduced:—*U. S. Ph.*, and others				
" **saccharated,** ⎫ —so-called,—see Iron, oxide, red, saccharated				
" **soluble** ..... ⎭				
" —**saccharated Salts** of,—see *reference* under Iron, saccharate.				
Iron and **Ammonium**, chloride, (so-called "Ammoniated Iron "), see Ammonium, chloride, with Ferric Chloride ..				
" and **do**., arsenicico-citrate, see Iron, arseniate *and* citrate, ammoniated...				
" and **do**.:—Citrate; Sulphate; Tartrate, —*all U. S. Ph.*,—see Iron, Sesqui-compounds: Ammonio-Ferric citrate;—sulphate;—tartrate.				
" and **Calcium**, lacto-phosphate, see Calcium, ferro-lacto-phosphate				
" and **Lead**, cyanide, so-called, see Lead, ferro-cyanide				
" and **Lithium**, salts, see "Lithium, ferro- —," etc......................				
" and **Mercury**, cyanide, so-called, see Mercury, ferro-cyanide				
" and **Potassium**, ferro-cyanide, (Potassium Ferri-ferro-cyanide; Soluble Prussian Blue), see Iron, cyanide, blue,—so-called,—soluble				
" and **do**., tartrate, *U. S. Ph.*,—(*Tartarated* [*Tartarized*] *Iron*,—NOT : "Iron-Tartar"!),—see Iron, *Sesqui*-compounds: Potassio-*Ferric* tartrate .... N.B.—*Compare, also:*—Iron, *Mono*-compounds: Potassio-*Ferrous* tartrate,—(*Ferrated Tartar; Iron-Tartar*);—etc.; etc.				
" and **Quinine**, citrate,—*U. S. Ph.* and other formulas, — see Quinine, ferri-citrate, etc..				

☞ **When ordering, specify : "MERCK'S"!**

	Containers incl.			
Iron and **Quinine**,—*other double salts* (than above),—see "Quinine, ferri- —," etc.				
" and **Strychnine**, citrate, *U. S. Ph.*, see Strychnine, ferri-citrate............				
" and **Zinc**, cyanide, so-called, see Zinc, ferro-cyanide...................				
Iron, Mono-compounds, (Ferro- double salts):				
Ammonio-Ferrous cyanide...............	lb. 2.50			
" sulphate, cryst..................	lb. .50			
Magnesio-Ferrous citrate	oz. .25			
" do., effervescent, yellow............	oz. .30			
" lactate.........................	oz. .50			
Mangano-Ferrous carbonate	oz. .35			
" do., saccharated...................	oz. .35			
" chloride..........................	oz. .40			
" citrate	oz. .30			
" cyanide...........................	oz. .30			
" iodide	oz. 1.00			
" lactate..........................	oz. .35			
" pyro-phosphate....................	oz. .40			
" sulphate	oz. .20			
Potassio-Ferrous citrate	oz. .35			
" cyanide, so-called, (Yellow Prussiate of Potassa),—see Potassium, ferrocyanide, *U. S. Ph.*, etc......				
" tartrate, (*Ferrated Tartar, Iron-Tartar;* —*not to be confounded with:* TARTARATED [TARTARIZED] IRON,—*which see, under:*— Iron, Sesqui-compounds: Potassio-Ferric tartrate, *U. S. Ph.*);—powder	lb. .75			
" " in globules, (so-called: Ir n Pellets, Steel Pellets)	lb. .85			
" " green.......................	lb. 2.00			
Sodio-Ferrous benzoate..................	oz. 1.00			
" citrate	oz. .35			
" cyanide, so-called, see Sodium, ferrocyanide........................				
Iron, Sesqui-compounds, (Ferrid- double salts):				
Aluminio-Ferric sulphate, see Alum, ferric				
Ammonio-Ferric arseniate and citrate, see Iron, arsen. *and* citr., ammoniated.				
" bromide	oz. .50			
" chloride, (so-called "Ammoniated Iron"), see Ammonium, chloride, with Ferric Chloride............				
" citrate, brown, in scales,—*Ferri et Ammonii citras, U.S.Ph.*	lb. 1.10			
" " green, in scales ...	lb. 1.40			
" cyanide...........................	oz. 1.75			
" oxalate, cryst.	lb. 1.50			
" sulphate,—*Ferri et Ammonii sulphas, U. S. Ph.*,—and Ph. G. I,— (Ammonio-Ferric Alum, Ammoniacal Iron-alum)..................	lb. .75			
" tartrate, (*Ammoniacal* Iron-Tartar, Ammonio-Ferric Tartar, Ferridammoniacal Tartar),—*Ferri et Ammonii tartras, U. S. Ph.*,—in scales..	lb. 1.50			
Calcio-Ferric cyanide, so-called, see Calcium, ferri-cyanide				
Mangano-Ferric phosphate, with Ammonium Citrate.......................				
Potassio-Ferric cyanide, so-called, (Red Prussiate of Potassa), see Potassium, ferrid-cyanide, etc..........				

☞ When ordering, specify: "**MERCK'S**"!

	Containers incl.			
Iron, Sesqui-compounds, (Ferrid- double salts),—*continued:*				
Potassio-Ferric oxalate, cryst.............	lb. 2.00			
" pyro-phosphate....................	oz. .75			
" sulphate, (Potassio-Ferric Alum, Potassic Iron-alum), pure...........	lb. .60			
" tartrate,—*Ferri et Potassii tartras, U. S. Ph.*,—(*Tartarated Iron, Tartarized Iron*),—brown, in scales.........	oz. .30			
N. B.—The above is *not to be confounded with:* FERRATED TARTAR; IRON-TARTAR,—*which see, under:*-Iron, Mono-compounds: Potassio - Ferrous tartrate, — powder; do. do., globules; do. do., green.				
Sodio-Ferric oxalate......................	oz. .30			
" pyro-phosphate....................	oz. .30			
" " in scales......................	oz. .35			
" tartrate, in scales	oz. .30			
Iron-Albumin, in scales ; and do., peptonized; and do., saccharated;—see Iron, albuminate, etc....................................				
N.B.—*Compare, also:* Iron, lactate .... ⎫ " phosphate .... ⎬ *albuminated.* " pyro-phosphate ... ⎭				
Iron Alum, see Alum, ferric...............				
" " **ammoniacal,** see Iron, Sesqui-compounds: Ammonio - ferric sulphate.....................				
" " **potassic,** see do., do.: Potassio-ferric sulphate................				
Iron Ethiops, see Iron, oxide, black.......				
Iron Pellets, so-called, see Iron, Mono-compounds : Potassio - Ferrous tartrate, in globules..................................				
Iron-Sugar (Ferruginated Sugar), [so-called "Saccharated Iron" or "Soluble Iron"], see Iron, oxide, red, saccharated..........				
N.B.—*Compare, also:* Iron, albuminate .... ⎫ " carbonate—(*U.S. Ph.;* etc.)— ⎪ " iodide—(*U. S. Ph.*)—....... ⎬ *saccharated.* " peptonized................ ⎪ " sulphate, Ferrous .... ⎪ " Mono - compounds : Manga- ⎪ no-Ferrous carbonate .... ⎭				
Iron - Tartar (*Ferrated* Tartar), see Iron, *Mono* - compounds: Potassio - *Ferrous* tartrate, etc.				
N. B.—*Compare, also:* Iron, *Sesqui*-compounds, Potassio-*Ferric* tartrate, *U. S. Ph.*, — (*Tartarated* [Tartarized] *Iron*).				
" **ammoniacal,** (*Ammonio-Ferric* Tartar; *Ferrid-ammoniacal* Tartar), see Iron, Sesqui-compounds: *Ammonio*-Ferric tartrate, *U. S. Ph.*................				
Iron Vitriol, (*Green Vitriol*), see Iron, sulphate, Ferrous:—*U.S.Ph.; do.* precipitated; *do.* exsiccated;—and other grades and forms				
Isatin	15 gr. 1.00			
Iso-butyl-aldehyd (Iso-butyr-aldehyd)....	15 gr. .50			
Iso-butyl-carbinol, see Alcohol, amylic, primary..................................				
Iso-Naphthol, see Naphthol, Beta-........				
Iso-propyl-benzene (-benzol), see Cumene.				
Iso-propyl-carbinol, see Alcohol, butylic, Iso-.....................................				

☞ When ordering, specify : **"MERCK'S"** !

Ivory-black, so-called, (Purified or Pure Bone-black), see Charcoal, animal, purified, *U. S. Ph.;* and do., pure..................

	Containers incl.		
Jaborine	15 gr. 4.00		
Jalapin — (identical with S<small>CAMMONIN</small>); — ["White Resin" of Fusiform Jalap]. — *The pure Glucoside* from Male (light, Orizaba) Jalap-root — Ipomœa orizabensis; *or* from Scammony-root	oz. 1.00		
N.B. — *See, also:* — Resins: Jalap, — brown: from the *light* Root.			
James's Antimonial Powder, (J.'s *Febrile* Powder), see Antimonial Powder, *U. S. Ph.*			
Jervine	15 gr. 4.00		
Juglandin	15 gr. .35		
Juice of Juniper-berries, inspissated, see Extracts: Juniper			
" of **Papaw (Carica papaya**—Melon-tree), —dry	⅛ oz. vls. oz. 2.00		
" of **Snails**, saccharated, see Helicina..			
Juices (Succi), *from fresh herbs*,—all according to *U. S. Ph. of* 1870:—			
Belladonna (Deadly Nightshade): leaves and young branches	lb. 1.00		
Conium (Hemlock): leaves	lb. 1.00		
Digitalis (Foxglove): leaves	lb. 1.00		
Hyoscyamus (Henbane): leaves and young branches	lb. 1.00		
Scoparius (Broom): tops	lb. 1.10		
Taraxacum (Dandelion): root	lb. 1.00		
Juniper-tar, see Oils, divers : Cade			
Kali, Kalium, and compounds, — see Potassa, etc.; and, Potassium, etc.			
Kamalin, cryst.	15 gr. .25		
Karlsbad Thermal Salt, — artificial; and, true, — see Salt, Karlsbad, etc., etc.; etc.			
Kefir (Kephir) **Fungi**	oz. 1.00		
Keratin (Corneous Substance, Horn-substance)	oz. .75		
Keratin, pepsinized; for coating Ileac pills, — acc. to Dr. Unna	oz. 6.00		
N.B. — *Ileac pills* are to pass the stomach undissolved, and develop their action only in the intestines.			
Kermes Mineral, see Antimony, sulphide, red, — so-called			
King's Yellow, see Arsenic, Yellow sulphide			
Kosin Merck, cryst., —(Cosin)	15 gr. 1.00		
Koussein Merck, amorphous, —(Coussein, Kossein; Brayerin)	⅛ oz. vls. oz. 6.00		
Kreatine, and Kreatinine, see Creatine, and Creatinine			
Kreosote, see Creasote			
Kresol, see Acid, cresylic			
Kreuznach Salt, ("Kreuznacher Mutterlaugensalz"), see Salt, Kreuznach			

☞ When ordering, specify : "**MERCK'S**" !

	Containers incl.
Lac Sulphuris purum, see Sulphur, precipitated, pure, *U. S. Ph.*	
Lacmoid, chem. pure, in scales;—an extremely sensitive substitute for Litmus....	⅙ oz.vls.oz. 3.00
Lacmus (*Chemically Pure* Litmus),—according to Wartha;—free from Lime and from the reddish colorifics soluble in Alcohol...	oz. 1.00
N.B.—*See, also:* Litmus, commercial.	
Lacto-Pepsin (miscalled "Lacto-peptine") [also called "Lactated Pepsin"], see Pepsin, Lacto-.	
Lactose (*Lactin*), see Milk-sugar.	
Lactucarium, Gallic, (Thridace), [Dried milk-juice of Garden Lettuce – *Lactuca sativa*],—in tablets	oz. .40
Lactucarium, Germanic, (the so-called "Lettuce-opium"),— (Dried milk-juice of Acrid Lettuce— *Lactuca virosa.*)	
first choice	oz. .60
' do., – II.	oz. .45
" " crumbs	oz. .40
" " fine powder	oz. .50
" " purified, — soft or dry, — see Extracts: Lactucarium	
Lactucin,—from Lactucarium	15 gr. 4.50
Lævulose (Levulose), see Fruit-sugar, I....	
Lamine Sulphate.—(*Lamine*—the Alkaloid of Blind-Nettle [*Lamium album*]—is a powerful hemostatic, adapted for subcutaneous application.)	
Lana Collodii, see Collodion Cotton	
Lanolin (Cholestearin Fat), in tins	lb. .80
" chem. pure, anhydrous	
Lantanin	
Lanthan (Lanthanum), metallic, powder...	15 gr.10.00
" chloride	15 gr. 1.00
" oxide	15 gr. 1.50
" sulphate	15 gr. 1.00
Lapis divinus, (Divine stone, Ophthalmic stone), so-called, see Copper, aluminated	
" **infernalis,** see Silver, nitrate, cryst.; and, molded;— *U. S. Ph.;* and, grey..	
Laudanum, see Tinctures: Opium; simple.	
" **Sydenham's,** see Tinctures: Opium,—saffronated	
Lead (Plumbum), double salts of, see "Lead and—" (below!)	
" metallic, pure, bars	lb. .65
" " " ribbon	lb. 1.00
" " " granulated,–free fr. Silver	lb. .35
" " chem. pure, powder	lb. 1.00
" acetate, mono-plumbic,— *U. S. Ph.*,— (Sugar of Lead — Saccharum plumbi [saturni]), chem. pure, cryst	lb. .50
" " do., pure, cryst	lb. .45
" " " purified, cryst	lb. .40
" acetate, basic (tri-plumbic, tri-basic), [Sub-acetate of Lead]	lb. 1.75
" " " —solution, *U. S. Ph.*, [Vinegar of Lead; "Goulard's Extract"], see under Solutions..	
" benzoate	oz. .65
" borate	oz. .30
" bromide	oz. .50
" carbolate, see Lead, phenate	
" carbonate, neutral, purified	lb. .50
" " " chem. pure	lb. 1.00

☞ When ordering, specify: **"MERCK'S"**!

		Containers incl.			
Lead, carbonate, basic, (oxy-carbonate; hydrico-carbonate), [White Lead],— Plumbi carbonas, U. S. Ph.					
" chloride, pure		lb. 1.00			
" " II		lb. .60			
" chromate, pure, fused		lb. 1.10			
" " " powder		lb. 1.10			
" cyanide		oz. .50			
" ferro-cyanide		oz. .25			
" formate, pure, dry		oz. .60			
" hydroxide (hydrate), mono-plumbic, [Mono-hydrated Prot-oxide of Lead], see Lead, oxide, mono-hydrated					
" hypo-phosphite		oz. .75			
" hypo-sulphite, see Lead, thio-sulphate.					
" iodide, powder,— U. S. Ph.		oz. .36			
" " cryst.		oz. .60			
" lactate		oz. .35			
" malate, pure		oz. 1.25			
" molybdate (molybdenate)		oz. 1.00			
" mono-chlor-acetate		oz. 5.00			
" nitrate		lb. .35			
" " pure,— U. S. Ph.		lb. .50			
" nitrite		oz. .50			
" oleate		oz. .25			
" oxalate		lb. 1.50			
" oxide (prot-oxide, mon-oxide; yellow oxide), anhydrous, fused,—[Litharge],—pure		lb. .70			
" " do., do., chem. pure,—U. S. Ph.		lb. 1.10			
" " mono-hydrated, (Mono-plumbic Hydroxide), pure		lb. 2.50			
" per-oxide (bin- [di-] oxide; brown oxide), —[Anhydrous Plumbic Acid],— (Puce [Brown] Lead)		lb. .60			
" " pure		lb. .85			
" phenate (phenylate, carbolate)		oz. .35			
" phosphate, pure		oz. .30			
" phosphite		oz. .50			
" rhodanide, see Lead, sulpho-cyanate					
" salicylate		oz. .75			
" silicate		oz. .25			
" sub-acetate, see Lead, acetate, basic					
" " solution, U. S. Ph.,—(Vinegar of Lead; "Goulard's Extract"),— see Solut's: Lead acetate, basic					
" sulphate, (Lead Vitriol)		lb. .40			
" " chem. pure		lb. .50			
" sulphide (sulphuret)		lb. 1.35			
" sulphite		lb. 1.50			
" sulpho-carbolate (phenol-sulphonate, sulpho-phenate)		oz. .30			
" sulpho-cyanate (thio-cyanate; rhodanide)		oz. .25			
" tannate, dry		oz. .30			
" tartrate		oz. .25			
" thio-cyanate, see Lead, sulpho-cyanate					
" thio-sulphate (formerly called "hyposulphite")		lb. .75			
" vanadate		15 gr. .75			
" wolframate (tungstate)		oz. 1.25			
Lead, puce (brown), see Lead, per-oxide; etc.					
" white, see Lead, carbonate, basic, U. S. Ph.					
Lead and Iron, cyanide, so-called, see Lead, ferro-cyanide					
" and Platinum, cyanide, see under Platinum double Cyanides					
" and Sodium, thio-sulphate (formerly called "hypo-sulphite")		oz. .50			

☞ When ordering, specify: "MERCK'S"!

	Containers incl.			
Lead, so-called Sugar of, see Lead, acetate, normal, *U. S. Ph.*				
" Vinegar of, ("Goulard's Extract"), see Solutions: Lead acetate, basic, *U.S.Ph.*				
" Vitriol of, see Lead, sulphate, etc.				
Leaves, Senna-, —free from resin, —see Senna, leaves, deresinated				
Lecithin	15 gr. 2.50			
Lemon-camphor, so-called, see Turpentine-oil, di-hydrochlorate				
Legumin (Vegetable Casein from legumes)	15 gr. .40			
Lepidine	oz. 1.00			
Leptandrin	oz. .50			
Leptandrin Merck, pure	oz. 2.50			
Lettuce-opium, so-called, see Lactucarium, Germanic, etc.				
Leucine, pure, (Amido-caproic Acid)	15 gr. 2.00			
" hydro-chlorate	15 gr. 2.00			
Leucoline (Leucol), synthetic, see Quinoline				
Leucotin, from Coto-bark	15 gr. .40			
Levulose (Lævulose), see Fruit-sugar, I.				
Libavius's Fuming Spirit, so-called, see Tin, tetra-chloride				
Lignite Tar, see Oils, divers: Lignite				
Lime (Calx), — *U. S. Ph.*, — (Pure Burnt Lime), [Dry Caustic Oxide of Calcium], — from marble	lb. .40			
Lime, antimonio - sulphurated (*stibiato*-sulphurated), [Antimonic Liver of Lime; Antimoniated (Stibiated) Calcic Liver of Sulphur; Calcic Liver of Antimony], (Calx Antimonii [Stibii] *cum Sulphure*), — [so-called "Antimonio-sulphide of Calcium"]	lb. .75			
Lime Hydrochlorate,—so-called,—see Calcium, chloride				
" Saccharate (bi - saccharate), — so-called,—see Calcium, saccharate				
Lime, sulphurated, — *U. S. Ph.*,—(Liver of Lime; Calcic Liver of Sulphur), [sometimes mis-called "Sulphide of Calcium"]	lb. .50			
Lime-water, see Solutions: Lime, *U. S. Ph.*				
Lipanin				
Liquid, Dutch, see Ethylene, chloride (bi-chloride)				
Liquid (Water-) Glass, see Potassium, silicate, etc.; and, Sodium, silicate, *U.S.Ph.*; etc.				
Liquor ammoniæ, (Liquor ammonii caustici), see Ammonia, Water of				
" ammonii caustici spirituosus Dzondii, see Ammonia, Spirit of				
" " acetatis, see Solutions: Ammonium acetate				
" anodynus martiatus, see Tinctures: Iron chloride, ethereal				
" seriparus, (Liquor ad serum lactis parandum), see Rennet Wine				
Liquores, *others than above*, see Solutions				
Litharge, pure; and, chem. pure;—see Lead, oxide, anhydrous, fused, pure ; and, chem. pure, *U. S. Ph.*				
Lithium, double and triple salts of, see "Lithium and —" (below!)				
' metallic	15 gr. 10.00			
" acetate	oz. .75			
" arseniate (arsenate)	oz. 1.25			
" benzoate,—*U. S. Ph.*	oz. .50			
" bi-borate	oz. .75			
" bi-carbonate, *so-called*, see Lithium, carbonate, bi-				
" bi-chromate	oz. .60			

☛ When ordering, specify: "MERCK'S"!

	Containers incl.
Lithium, boro-citrate.....................	oz. .75
" bromide,—*U. S. Ph.*.................	oz. .38
" carbolate, see Lithium, phenate	
" carbonate...........................	oz. .36
" " chem. pure,—*U. S. Ph.* & Ph.G. II	oz. .38
" " effervescing..................	oz. .30
" " bi-, — so-called, — is only Lithium carbonate!	
" chloride............................	oz. .45
" chromate, bi-, see Lithium, bi-chromate	
" citrate, cryst.,—Ph. Brit. new.........	oz. .36
" " powder,—*U. S. Ph.*............	oz. .35
" " effervescing....................	oz. .30
" ferro-benzoate......................	oz. 1.00
" " -citrate	oz. 1.00
" hippurate	oz. 2.50
" ichthyol - sulphonate, see under Ichthyol preparations.................	
" iodide	oz. .67
" lactate..............................	oz. .75
" nitrate..............................	oz. .75
" oxalate..............................	oz. 1.00
" oxide, caustic.......................	oz. 1.40
" phenate (phenylate, carbolate)	oz. 1.00
" phosphate...........................	oz. 1.25
" salicylate,—*U. S. Ph.*,—chem. pure, perf. white................................	oz. .49
" succinate............................	oz. 1.00
" sulphate, cryst......................	oz. .45
" sulpho-carbolate (phenol-sulphonate, sulpho-phenate)....................	oz. .60
" sulpho-ichthyolate, see under Ichthyol prep.	
" tartrate	oz. .75
" urate................................	oz. 2.00
" valerianate	oz. 1.00
Lithium and Iron, benzoate; and, citrate;—see "Lithium, ferro- —," etc.; etc...	
" and Potassium, tartrate	oz. 1.75
" and Sodium, benzoate...............	oz. .65
" " " salicylate	oz. .60
Lithium, Platinum, and Potassium, cyanuret, see under Platinum triple Cyanides...................................	
Litmus, chem. pure, see Lacmus..........	.
" commercial.....................	
Litmus Paper, red or blue, see under Paper	
Liver of Antimony, — (sometimes called: "Unwashed Brown Oxide of Antimony"), — see Potassa, antimonio-sulphurated, *crude*	
" " " calcic, (Antimonic Liver of Lime), see Lime, antimonio-sulphurated..............	
" of Lime, (*Calcic Liver of Sulphur*), see Lime, sulphurated, *U. S. Ph.*	
" " " antimonic, (Calcic Liver of Antimony), see Lime, antimonio-sulphurated	
" of Sulphur, (*Potassic L. of S.*), see Potassa, sulphurated, *U. S. Ph.*; and other grades	
" " " calcic, see Lime, sulphurated, *U. S. Ph.*	
" " " " —antimoniated (*stibiated*), [Antimonic Liver of Lime], see Lime, antimonio-sulphurated	
" " " sodic, see Soda, sulphurated, etc.	
Lobeline, sulphate.........................	15 gr. 2.50

☞ When ordering, specify: "MERCK'S"!

	Containers incl.		
Lunar Caustic, see Silver, nitrate, molded, *U. S. Ph.;* and, grey; and, do., pencils			
" " **mitigated** (toughened), see Silver, nitrate, diluted, *U. S. Ph.;* and other strengths..........			
" **Nitre,** see Silver, nitrate, cryst., *U. S. Ph.*			
Lupulin, purified, new crop	lb. .50		
Luteoline, see under Aniline and Phenol Dyes : Yellow............			
Lutidine (Di-methyl-pyridine).............	oz. 2.50		
Lycoctonine	15 gr. 2.50		

☞ When ordering, specify : "**MERCK'S**"!

	Containers incl.			
Madagascar Sugar, see Melampyrit.....				
Magdala Red, see under Aniline and Phenol Dyes: Red............................				
Magistery of Bismuth, see **Bismuth, sub-nitrate,** chem. pure, *U. S. Ph.*............				
" of **Sulphur,** see Sulphur, precipitated, pure, *U. S. Ph.*...................				
Magnesia, *U. S. Ph.*, — light, — (Light Calcined Magnesia—Magnesia usta levis), —see Magnesium, oxide, light.......				
" alba, so-called, (Magnesia hydrico-carbonica), see Magnesium, carbonate, light, *U. S. Ph.*....................				
" **ponderosa,** *U. S. Ph.*, (Heavy Calcined Magnesia), see Magnesium, oxide, heavy........................				
Magnesia Hydrate, moist, see Magnesium, hydroxide, moist.....................				
Magnesia, ricinated, see Magnesium, ricinate...............................				
Magnesium, double salts of, see "Magnesium and —" (below!)...............				
" metallic, bars......................	oz. 1.00			
" " wire or ribbon................	oz. 1.00			
" " powder	oz. 1.00			
" acetate...........................	oz. .20			
" æthylo-sulphate, see Magn., eth.-sulph.				
" benzoate..........................	oz. .40			
" bi-phosphate, so-called, see Magnesium, phosphate, acid...................				
" bi-sulphate.........................	lb. 2.00			
" borate	oz. .25			
" boro-citrate, powder.................	oz. .22			
" " scales...................	oz. .30			
" bromide............................	oz. .42			
" carbonate, heavy (cryst.) [neutral].....	lb. 1.25			
" " light (so-called "amorphous") [basic],—(sub-carbonate),—[so-called "Magnesia alba"; Magnesia hydrico-carbonica],—*Magnesii carbonas, U. S. Ph.*	lb. .50			
" chloride, crude	lb. .30			
" " pure, cryst.....................	lb. .40			
" " chem. pure, cryst...............	lb. .50			
" " " " fused...............	lb. .75			
" citrate, soluble	lb. 1.50			
" " in scales	oz. .40			
" " effervescent,—Ph. G. II,—(*Pulvis aërophorus* cum Magnesia citrica)......................	lb. 1.25			
" " effervescent, granulated,—*U. S. Ph.*,—(*Granella aërophora* cum Magnesia citrica)	lb. .75			
" ergotate, see Magnesium, sclerotate ...	15 gr. .50			
" ethylo-sulphate (sulpho-vinate)	oz. .35			
" formate	oz. .50			
" hydroxide, (Magnesia Hydrate), moist, pultaceous, [Magnesia hydrica pultiformis],—according to the Table of Re-agents of Ph. G. II..............	lb. .75			
" **hypo-phosphite,** chem. pure, cryst........	oz. .35			
" hypo-sulphite, see Magnesium, thio-sulphate.............................				
" iodide..............................	oz. .70			
" lactate, pure........................	oz. .35			
" lacto-phosphate (phospho-lactate)	oz. .35			
" malate	oz. 1.50			
" nitrate, pure........................	lb. 1.00			
" oxalate............................	lb. 1.50			

☞ **When ordering, specify: "MERCK'S"!**

		Containers incl.	
Magnesium, oxide, light, (Light Calcined Magnesia — Magnesia usta levis), — *Magnesia, U. S. Ph.*	lb.	.55	
" " heavy, (Heavy Calcined Magnesia), — *Magnesia ponderosa, U. S. Ph.*	lb.	.75	
" " hydrated, moist, see Magnesium, hydroxide, moist............			
" phosphate, acid, (so-called "bi-phosphate")	oz.	.35	
" " neutral, (Tri-magnesic ortho-Phosphate), pure.............	oz.	.19	
" " do., II	oz.	.18	
" phospho-lactate, see Magnesium, lacto-phosphate			
" rhodanide, see Magnesium, sulpho-cyanate			
" ricinate, (Magnesia-and-Castor-oil Soap —Sapo ricini magnesicus), [Ricinated Magnesia]........................	lb.	1.75	
" salicylate, cryst., — easily soluble. — (A mild succedaneum for Bismuth Salicylate.).........................	oz.	.55	
" sclerotate (ergotate)..................	15 gr.	.50	
" silicate............................	oz.	.35	
" succinate...........................	oz.	.60	
" sulphate, (Epsom Salt—Sal amarum), cryst., perfectly colorless	lb.	.30	
" " dry, perfectly white.............	lb.	.35	
" " chem. pure, cryst., — *U. S. Ph.*...	lb.	.35	
" " " " exsiccated..........	lb.	.35	
" " bi-, see Magnesium, bi-sulphate..			
" sulphite, — *U. S. Ph.*.................	lb.	.80	
" sulpho-carbolate (sulpho-phenate, phenol-sulphonate)....................	oz.	.30	
" sulpho-cyanate (thio-cyanate; rhodanide)...............................	oz.	.30	
" sulpho-vinate, see Magnesium, ethylo-sulphate			
" tartrate, — according to Rademacher ...	oz.	.35	
" thio-cyanate, see Magn., sulpho-cyanate			
" thio-sulphate (formerly called "hypo-sulphite")	oz.	.25	
" urate ....,,..	oz.	1.00	
" valerianate	oz.	1.00	
Magnesium and **Ammonium,** arseniate (arsenate)	lb.	2.00	
" and do., chloride — [Mg Cl$_2$. N H$_4$ Cl. 6 H$_2$O]. — (Used for preparing the Magnesia mixture for the determination of Phosphoric Acid.)..................			
" and do., phosphate..................	lb.	2.00	
" " " sulphate...................	lb.	.60	
" and **Iron,** salts, see under Iron, Mono-compounds........................			
" and **Platinum,** cyanide, see under Platinum double Cyanides.............			
" and **Sodium,** salts, see Sod. and Magn.			
Magnetic Oxide, see Iron, oxide, black....			
Magnus's "Green Salt," see Platinum double Chlorides: Platinum tetr-amine and Platinum, bi-chloride....................			
Malachite, blue, artificial, see Copper, carbonate, blue......................			
" (*Green Malachite*), artificial, see Copper, carbonate, green................			
Malachite Green, (*not* in any manner related to *Green Malachite!*), see under Aniline and Phenol Dyes: Green..................			
Maltin, see **Diastase of Malt**...............			

☞ When ordering, specify: "**MERCK'S**"!

	Containers incl.			
Manchester Yellow, see under Aniline and Phenol Dyes: Yellow..................				
Manganese (Manganum), double salts of, see "Manganese and —" (below!)...				
" metallic.............................	15 gr. .15			
" acetate.............................	oz. .25			
" arseniate (arsenate), pure............	oz. .45			
" benzoate............................	oz. 1.00			
" bin-oxide, see Manganese, per-oxide, *artificial;* — *also:* Manganese, oxide, black, *U. S. Ph.*....................				
" bi-silicate, see Manganese, silicate.....				
" borate.—[A paint-drier (siccative).]....	lb. .45			
" bromide.............................	oz. .62			
" carbonate, Manganous, chem. pure....	lb. 2.00			
" chloride, Manganous, pure, cryst......	lb. 1.00			
" " " " fused.....	oz. .40			
" " " crude..........	lb. .40			
" citrate..............................	oz. .50			
" di-oxide, see Manganese, per-oxide, *artificial;—also:* Manganese, oxide, black, *U. S. Ph.*......................				
" **hypo-phosphite, chem. pure, cryst.**........	oz. .35			
" hypo-sulphate.......................	oz. 1.00			
" iodide...............................	oz. .75			
" lactate..............................	oz. .45			
" lacto-phosphate (phospho-lactate).....	oz. 1.00			
" nitrate, pure........................	oz. .30			
" oleate...............................	oz. .35			
" oxalate..............................	oz. .30			
" oxide, sesqui-, (Manganic oxide), anhydrous, pure.............	lb. 2.00			
" " " hydrated.................	lb. .75			
" " " black,—*U. S. Ph.*,—(*Native* Peroxide [Bin-oxide, Di-oxide] of Manganese), — [at least 66% MnO_2]; — (Black Manganese; also called "Pyrolusite")......	lb. 2.00			
" " do., purified, see Manganese, peroxide.............................				
" per-oxide (di-oxide), *artificial*, pure,— [abt. 90% MnO_2]; — (*Purified Black Oxide* of Manganese; *Purified* Black Manganese).........................	lb. 2.00			
" phosphate, Manganous, pure..........	oz. .45			
" phospho-lactate, see Manganese, lacto-phosphate...........................				
" salicylate............................	oz. 1.50			
" sesqui-oxide, see Manganese, oxide, sesqui-; etc.........................				
" silicate (bi-silicate).—[Used in enameling.]................................	oz. .40			
" succinate............................	oz. 1.00			
" sulphate, Manganous, crude..........	lb. .50			
" " do., pure, cryst.,—*U. S. Ph.* and *Ph. G. II*........	lb. .80			
" " " exsiccated............	lb. 2.00			
" sulphite.............................	lb. 1.75			
" sulpho-carbolate (phenol-sulphonate, sulpho-phenate)....................	oz. .50			
" tannate.............................	oz. .55			
" tartrate.............................	oz. .55			
" valerianate.........................	oz. 1.50			
Manganese, black; and: do., do., purified; —see Manganese, oxide, black,—*U. S. Ph.*; and: do., per-oxide, *artificial*.........				
Manganese and Iron, salts, see under Iron, Mono-compounds; and under Iron, Sesqui-compounds..................				
" and **Zinc,** chloride, see Z. and M., chl.				

☞ When ordering, specify: "**MERCK'S**"!

	Containers incl.	
Manna-sugar, ⎱ (Mannitol, Mannol; Fraxin- **Mannit**......⎰ in; Granatin;—formerly also called "Punicin")..	lb. 2.50	
" recrystallized from Alcohol............	oz. .40	
Martius Yellow, see under Aniline and Phenol Dyes: Yellow..................		
Mass (Pill-mass), mercurial, [Mass of Mercury—*Massa hydrargyri, U. S. Ph.*;— Blue Mass]......................	lb. 2.50	
" Vallet's, (Mass of Carbonate of Iron — *Massa ferri carbonatis, U. S. Ph.*;— Massa ferrata).....................	lb. .75	
Meat-sugar, see Inosit..................		
Meconin (Opianyl).......................	15 gr. 1.00	
Melampyrit (Melampyrin; Dulcit, Dulcin, Dulcol, Dulcose, Dulcitol; Evonymit) [*Madagascar Sugar*], cryst.	oz. 2.50	
Melanin...............................	$^{15}/_{100}$gr. 1.00	
Menthol (Peppermint-camphor), Japanese, cryst., dry,—in original 5-lb. tins, or in broken packages................	lb. 3.00	
" recrystallized, chem. pure	lb. 4.00	
" benzoated...........................	oz. 1.50	
Mercaptan, ethylic, (Ethyl Hydrosulphide [Sulphydrate]; Ethylic Thio-alcohol).....	15 gr. .35	
Mercur-ammonium, chloride, see Mercury, ammoniated, so-called, *U. S. Ph.*,— infusible........................		
" **-di-ammonium,** chloride, see do., do., do., fusible.......................		
" **-di-benzene** (*Di-phenyl-mercury*).—See remark relating to this *non-medicinal*, extremely poisonous *metallo-organic* compound,—under: "Mercury, diphenate"; with which the former is sometimes erroneously confounded.		
" **-thymol,** (Thymol-Mercury), acetate,— [Thymol-acetate of Mercury]		
Mercurial Ethiops, see Mercury, sulphide, black,—*so-called*.....................		
Mercury (Mercurius; Hydrargyrum), double salts of, see "Mercury and —" (below!)..................................		
" metallic,—*U. S. Ph.*...................	lb. .90	
" " chem. pure	lb. 1.05	
" acetate, Mercurous [Suboxide salt].....	oz. .40	
" " Mercuric [Peroxide salt].........	oz. .35	
" albuminated, fluid, —so-called, —see Mercury, bi-chloride, albuminated, etc. N. B.—*See, also*: Mercury, bi-chloride, albumino-saccharated, dry.		
" ammoniated, so-called, -(*amidato-bichloride*),—*U. S. Ph.* and Ph. G. II.,—*infusible*............ (Ammonio-chloride of Mercury; Mercurammonium Chloride;— Infusible *White* Precipitate)	lb. 1.50	
" do., do., *fusible*,—Ph. Neerl.,—(Mercur-di-ammonium Chloride; Fusible *White* Precipitate)........................	lb. 1.50	
N. B.—*The above two* preparations should *not be confounded* with the *following*:—		
" ammoniated *Nitrate* of, (*Black* Precipitate), see Mercury, oxide, black,—so-called.............................		
" antimonio-sulphide, (Antimonial Ethiops), [Black Sulphides of Antimony and Mercury; Mercurous Sulphide with Antimonious Sulphide]	lb. 1.25	
" arseniate (arsenate)	oz. .40	

☞ When ordering, specify: "**MERCK'S**"!

	Containers incl.			
Mercury, arsenite	oz. .60			
" arsenio-iodide, (Bin-iodide of Mercury with Ter-iodide of Arsenic)	oz. 1.00			
N. B.—*Solution of above double salt*, (Solution of Arsenic and Mercury Iodides, *U. S. Ph.*), [Donovan's Solution], see under Solutions.				
" benzoate	oz. .60			
" bi-bromide	oz. .45			
" bi-chloride,–*called* "corrosive chloride"! —(per-chloride), [Corrosive Sublimate], cryst. { *Hydrargyri chloridum corrosivum, U.S. Ph.* }	lb. 1.10			
" " powder	lb. 1.25			
" " recrystallized	lb. 1.50			
" " albuminated, (so-called "Albuminated Mercury"), *fluid*,—acc. to Bamberger, [Liquor hydrargyri albuminati B.];—containing 1% of Corrosive Sublimate.	oz. .35			
" " albumino-saccharated (saccharoalbuminated), *dry*, — acc. to Schneider, — containing 0.4% of Corrosive Sublimate.—[Used for wound-dressing, it furnishes a *constant* source of Hg Cl$_2$,—which salt is *gradually* dissolved-out by the serum secretion.]				
" " carbamidated (ureated), [Corrosive Sublimate with Urea], (so-called "Carbamidated" or "Ureated Mercury")	oz. 1.00			
" " peptonized, (so-called "Peptonized Mercury"), liquid,— [1% of Sublimate]				
" " " dry,—[10% of Sublimate]	oz. .50			
" bin-iodide (per-iodide) [red iodide], (Mercuric Iodide),—*Hydrargyri iodidum rubrum, U. S. Ph.*	oz. .34			
" " with Arsenic Ter-iodide, see Mercury, arsenio-iodide				
" bi-sulphate,–improperly so-called,–see Mercury, sulphate, Mercuric, *neutral*.				
" borate, Mercuric [Peroxide salt]	oz. .50			
" bromide	oz. .45			
" bi-, see Mercury, bi-bromide				
" carbamidated, — so-called, — see Mercury, *bi-chloride*, carbamidated				
" **carbolate,**—acc. to Dr. K. Schadeck,—see **Mercury, phenate**				
" carbolate, di-, see Mercury, di-phenate.				
" carbonate, Mercurous [Suboxide salt]	oz. .50			
" chloride, — *called* "mild chloride"!— (proto- or mono-chloride), [Calomel], (Hydrargyri chloridum *mite*),—sublimed,—in lumps	lb. 1.50			
" " do. "do. do.,"—sublimed,–levigated (washed)	lb. 1.50			
" " " " " condensed by steam	lb. 1.50			
" " " " " *U. S. Ph.*, — precipitated; by wet process	lb. 1.50			
" chloride, bi-	see Mercury, *bi-chloride, U. S. Ph.;* etc.			
" " corrosive,				
" " mild, see Mercury, chloride,–*called* "mild chloride"!–*U. S. Ph.;* etc.				
" chloro-iodide	oz. .50			
" chromate	oz. .40			
" citrate,—insoluble in Water and in Alcohol	oz. .50			

☞ **When ordering, specify: "MERCK'S"!**

	Containers incl.

Mercury, cyanide, cryst.,–*U. S. Ph.*–(Lately, a powerful specific in Diphtheria!) .. oz. .40
" di-phenate (di-phenylate, di-carbolate), = $Hg(C_6H_5O)_2$................. 15 gr. .50
    N. B.—The above medicinal substance (as also the simple Mercury Phenate), is *not to be confounded* — as some professional journals have done—with the destructively toxical, and non-medicinal, DI-PHENYL-MERCURY (Mercur-di-benzene) = $Hg(C_6H_5)_2$!
" ferro-cyanide, pure................. oz. .50
" **form-amidated, solution,**—[1% Per-oxide] lb. 1.00
"   " —[10% "] oz. .30
" **glyco-cholate, solution,**—[1% "] oz. .50
" gynocardate,—extract consistency.... oz. 1.50
" Hahnemann's soluble, see Mercury, oxide, black,—so-called............
" iodide, green ("yellow"), [prot-iodide], (Mercurous Iodide), — *Hydrargyri iodidum viride, U. S. Ph.*... oz. .31
"   " bin- (per-).... ⎫ see Mercury, bin-
"   " red, *U. S. Ph.*, ⎭ iodide........
"   " do., with Arsenic Ter-iodide, see Mercury, arsenio-iodide........
"   " sesqui-, see Merc., sesqui-iodide
" lactate............................. oz. 1.00
" mercaptide......................... 15 gr. .50
" methylo-chloride................... 15 gr. .50
" nitrate, Mercuric [Peroxide salt]...... oz. .25
"   " Mercurous [Suboxide salt], normal, cryst. .... oz. .25
"   " " basic, (Sub-nitrate of Mercury), [Nitric Turpeth] .. oz. .25
"   " ammoniated, (*Black* Precipitate), see Mercury, oxide, black,—so-called.....................
    N. B.—The above preparation should *not be confounded* with the so-called "Ammoniated Mercury," *U. S. Ph.*, etc., (*White* Precipitate);—*which see also!*
" oleate,—[15% Per-oxide]............ oz. .30
"   " —[10% "]............ oz. .25
" oxalate, Mercurous [Suboxide salt].... oz. .50
"   " Mercuric [Peroxide salt]........ oz. .55
" oxide, black, — so - called, — (Hahnemann's Soluble Mercury; Ammoniated *Nitrate* of Mercury), —[Black Precipitate] .... oz. .30
"   " red,—*U. S. Ph.*,—(Mercuric oxide; per-oxide,—by *dry* process),—[Red Precipitate].. lb. 1.60
"   " " —levigated .... lb. 1.75
"   " yellow, — *U. S. Ph.*, — (Mercuric oxide; per-oxide,—by *wet* process),—[Yellow Precipitate]..... oz. .18
" oxy-cyanide.—(Succedaneum for Mercury *bi-chloride;*—more powerful as a disinfectant; and better tolerated as a medicine.)............................
" oxy-sulphate, (Yellow Sub-sulphate of Mercury, *U. S. Ph.*), see Mercury, sulphate, Mercuric, *basic*.............
" palmitate,—[10% Per-oxide] .... oz. .35
" peptonized,—so-called,—liquid and dry, —see Mercury, *bi-chloride*, peptonized, etc.; etc. ....

☞ **When ordering, specify : "MERCK'S"!**

	Containers incl.			
Mercury, per-oxide, by *dry* process, see Mercury, oxide, red, *U. S. Ph.;* and: do., do., do., levigated..................				
" do., by *wet* process, see Mercury, oxide, yellow, *U. S. Ph.*...............				
" phenate (phenylate, carbolate),—according to Dr. K. Schadeck............	oz. 1.00			
N.B.—*Compare, also,* remark under Mercury, di-phenate.				
" phenate, di-, see Mercury, di-phenate..				
" phosphate, Mercuric [Peroxide salt]...	oz. .45			
" " Mercurous [Suboxide salt]	oz. .45			
" precipitate, black, (Hahnemann's Soluble Mercury), see Mercury, oxide, black,—so-called.........				
" " red, see Mercury, oxide, red, *U. S. Ph.;* and: do., do., do., levigated......................				
" " white, infusible, see Mercury, ammoniated, so-called, *U. S. Ph.,*—infusible......				
" " do., fusible, see do., do., do., fusible				
" " yellow, see Mercury, oxide, yellow, *U. S. Ph.*...............				
" rhodanide, see Mercury, sulpho-cyanate				
" saccharo-albuminated *Bi-chloride* of,—dry,—see Mercury, bi-chloride, albumino-saccharated, etc..............				
" salicylate. — (A new favored by recent syphilidologists.)...................	oz. 1.00			
" santoninate (*not* santonate!), Mercurous [Suboxide salt]	oz. 1.00			
" sesqui-iodide, (Mercuro-mercuric iodide)	oz. 1.00			
" soluble, Hahnemann's, see Mercury, oxide, black,—so-called				
" stearate............................	oz. .40			
" stibiato-sulphide, see Mercury, antimonio-sulphide.....................				
" sub-nitrate, see Mercury, nitrate, Mercurous, basic......................				
" sub-sulphate, yellow, *U. S. Ph.*, see Mercury, sulphate, Mercuric, *basic*..				
" sulphate, Mercuric [Peroxide salt], *neutral,*—(Per-sulphate of Mercury; sometimes improperly called "Bi-sulphate")........	lb. 1.00			
" " Mercuric, *basic,*(Turpeth Mineral), [Oxy-mercuric sulphate ; Oxysulphate of Mercury];—Yellow Sub-sulphate of Merc'y, *U.S.Ph.*	lb. 1.40			
" " Mercurous [Suboxide salt]	lb. 1.50			
" sulphide (sulphuret), black,—*so-called;* —[Mercurous sulphide, *with excess of Sulphur !*]; —*formerly:* *U. S. Ph.;*—(Ethiops Mineral, Mercurial Ethiops)	lb. .90			
" " red (Mercuric),—*U. S. Ph.,*—(Best Artificial Cinnabar; Vermilion).	lb. 1.30			
" sulphite, Mercuric [Peroxide salt], neutral............................				
" sulpho-cyanate (thio-cyanate; rhodanide)............................	oz. .35			
" **tannate, Mercurous** (Suboxide salt],—containing 50% of Mercury............	oz. .48			
" tartrate	oz. .40			
" thio-cyanate, see Mercury, sulpho-cyanate				
" thymol-acetate, see Mercur-Thymol, ac.				
" ureated (carbamidated),—so-called,—see Merc., *bi-chloride,* carbamidated ..				

☞ **When ordering, specify : "MERCK'S" !**

	Containers incl.		
Mercury, di-Phenyl-.—See *remark* under Mercury, di-phenate.			
Mercury and **Ammonium**, chloride, infusible, — Ph. G. II, — (Ammoniochloride of M., Amidato-bichloride of M., Mercur-ammonium chloride; Infusible White Precipitate),—see Mercury, ammoniated, so-called, *U. S. Ph.*,—*infusible*	.		
" and do., do., fusible, (Mercur-*di*-ammonium chloride; Fusible White Precipitate),— Ph. Neerl.,—see do., do., do., *fusible*			
" and do., sulphate, (*Tetra*-mercur-*di*-ammonium sulphate; Di-mercur-ammonium *basic* sulphate), [Ammoniacal Turpeth]	lb. 2.00		
" and **Antimony** Sulphides (Black Sulphides [Sulphurets]), see Mercury, antimonio-sulphide			
" and **Arsenic** Iodides, see Mercury, arsenio-iodide			
" and do. do., *solution; U. S. Ph.*, (Donovan's Solution), see Solutions : Arsenic and Mercury Iodides			
" and **Iron**, cyanide, so-called, see Mercury, ferro-cyanide			
" and **Potassium**, cyanide	oz. .65		
" " " iodide	oz. .75		
" " " tartrate	oz. .45		
Mercury Amalgams: of Sodium; of Tin and Zinc; and of Zinc;—see: Sodium Amalgam ; Zinc Amalgam ; Zinc and Tin, Amalgam			
Mercury with **Chalk,**—[1 part of Purified Mercury: 2 of Prepared Chalk]	lb. 1.25		
Mesitylene, chem. pure	15 gr. .40		
meta-Chloral, see Chloral, meta-			
meta-Di-amido-benzene (-benzol), meta-Phenylene-di-amine], hydrochlorate, see Di-amido-benzene, meta-, etc.			
meta-Di-oxy-toluene, *symm.*, see Orcin			
meta-Nitro-aniline, see Nitro-aniline, meta-			
Metal, fusible,—acc. to Rose	oz. 1.00		
" " " Wood	oz. 1.00		
Methol, see Alcohol, methylic			
Meth-oxy-Caffeine, see Methyl-oxy-Caff.			
Methyl, acetate	oz. .50		
" benzoate, (so-called "Essence of Niobe")	oz. .60		
" bi-chloride,—acc. to Richardson	oz. .75		
" cyanide, (Cyano-methyl), [Aceto-nitrile]	oz. 5.00		
" butyrate	oz. 2.00		
" formate	oz. 1.00		
" iodide, (Mono-iod-methane)	oz. 1.00		
" nitrate	oz. 1.00		
" oxalate	oz. 1.00		
" oxide, hydrated, see Alcohol, methylic			
" phenate, see Anisol			
" salicylate, (Mono-methylic Ether of Salicylic Acid), [so-called " Methyl-salicylic Acid," or "Gaultheric Acid").—The principal constituent of Wintergreen Oil	oz. .65		
" sebacylate	oz. 2.00		
Methyl Chloroform, (Di-chloride of Mono-chlorethylidene)	oz. 1.00		
Methylal	oz. 2.50		
Methyl-amine (Amido-methane), chloride	oz. 3.00		
Methyl-aniline	lb. 2.00		
Methyl-benzene (-benzol), see Toluene			
Methyl-glycocoll [-glycocine], see Sarcosine			

☞ When ordering, specify : "**MERCK'S** " !

	Containers incl.
Methyl-oxy-Caffeine (Meth-oxy-Caffeine).	15 gr. .75
Methyl-propyl-benzene (-benzol), para-, see Cymene.................	
Methyl-Strychnine..................	15 gr. 5.00
" hydro-iodate (hydriodate), cryst.......	15 gr. 2.00
Methylene Chloride (Bi-chloride) Merck, chem. pure,—[Di-chlor-methane]...............	oz. .60
Methylene-proto-catechu-aldehyd, see Piperonal, chem. pure.................	
Mezerein, see Extracts: Mezereon; alco., etc.	
Microcosmic Salt, see Sodium and Ammonium, phosphate................	
Milk-sugar (Saccharum lactis; Lactose, Lactin), cryst................	lb. .50
" powder................	lb. .50
" —U. S. Ph.,—recrystallized..........	- lb. .65
Milk of Sulphur, pure, see Sulphur, precipitated, pure, U. S. Ph.................	
Mindererus's Spirit, so-called, see Solutions: Ammonium acetate............	
Mineral Chameleon, (Chameleon Mineral), see Potassium, manganate............	
Mineral, Cobaltum-, so-called,—(so-called "Metallic" Arsenic), — see Arsenic, cryst................	
" Ethiops-, see Mercury, sulphide, black, —so-called................	
" Kermes-, see Antimony, sulphide, red, —so-called................	
" Turpeth-, see Mercury, sulphate, Mercuric, basic,—U. S. Ph.............	
Mirbane Essence, (Mirbane Oil),-so-called, —see Nitro-benzene................	
Mollin, pure..................	lb. 1.00
Mollin Ointments,—with:—	
Acid, carbolic,—3-5%..................	lb. 1.25
" salicylic,—3-5%..................	lb. 1.25
" tannic,—5%..................	lb. 1.25
Balsam, Peru-, —10%..................	lb. 1.50
Birch tar, (Pix betulæ),—10-20%........	lb. 1.25
Creolin,—1-2%.—(According to Prof. Dr. Esmarch, a Creolin ointment is preferable, as a gynecological lubricant, to a Corrosive-Sublimate preparation.)......	lb. 1.50
Chrys-arobin,—5%..................	lb. 1.50
Ichthyol,—10-50%..................	lb. 2.25
Iodoform,—10%..................	lb. 2.50
Mercury, "ammoniated," (White Precipitate),—10%................	lb. 1.50
" bi-chloride, (Corrosive Sublimate),—1%	lb. 1.50
" metallic,—(Blue Ointment),—33⅓%....	lb. 1.75
" " —(do. do.),—50%........	lb. 2.00
" red oxide, (Red Precipitate),—5%.....	lb. 1.50
Naphthalene,—10%..................	lb. 1.25
Naphthol, Beta-, —5%..................	lb. 1.15
Potassium Iodide,—10%................	lb. 2.25
Resorcin,—10%..................	lb. 1.25
Sozo-iodole................	lb. 2.00
Storax (Styrax),—10%..................	lb. 1.50
Sulphur,—30-50%..................	lb. 1.25
Thymol,—5%..................	lb. 1.75
Molybdenum (Molybdænium), metallic....	15 gr. .50
" oxide, pure................	oz. 1.00
Mono-brom-benzene (-benzol), [Bromated Benzene (Benzol)], (Phenyl Bromide)......	oz. 1.00
Mono-brom-ethane, see Ether, hydrobromic	
Mono-brom-naphthalene, Alpha-, see Naphthalene, Alpha-mono-bromated......	
Mono-brom-phenyl-acet-amide, see Brom-phenyl-acet-amide, mono-..........	

☞ When ordering, specify: "MERCK'S"!

	Containers incl.			
Mono-chlor-benzene (-benzol), [Chlorated Benzene (Benzol)], (Phenyl Chloride).....	oz. 1.00			
Mono-chlor-ethylene Di-chloride, (Hyper-chlor-acetyl)...............................	oz. 1.00			
Mono-chlor-ethylidene Di-chloride, see Methyl Chloroform...............................				
Mono-chlor-hydrin.....................	oz. 1.50			
Mono-chlor-toluene (-toluol)............				
Mono-iod-benzene (-benzol), [Iodated Benzene (Benzol)], (Phenyl Iodide)..........	oz. 5.00			
Mono-iod-ethane, see Ether, hydro-iodic........				
Mono-iod-methane, see Methyl, iodide...				
Mono-methyl-catechol — (Absolute [Medicinal] Guaiacol), - see Guaiacol, chem. pure.......				
Monsel's Salt, see Iron, sub-sulphate.....				
" Solution, see Solutions : Iron sub-sulphate, U. S. Ph.................				
Mordant ("Preparing-") Salt, see Sodium, stannate.....................				
Morphine (Morphia, "Morphium"), pure, cryst.,—*Morphina, U. S. Ph*........	⅛ oz.vls.oz. 4.50			
" pure, precipitated....................	⅛ oz.vls.oz. 4.35			
" acetate,—*U. S. Ph*.	⅛ oz.vls.oz. 2.90			
" arseniate (arsenate)..................	⅛ oz.vls.oz. 6.00			
" asparagate...........................	⅛ oz.vls.oz. 7.00			
" benzoate	⅛ oz.vls.oz. 5.50			
" bi-meconate, see Morphine, meconate..				
" borate	⅛ oz.vls.oz. 6.00			
" camphorate	⅛ oz.vls.oz. 7.00			
" citrate	⅛ oz.vls.oz. 6.00			
" ferro-hydrocyanate..................	⅛ oz.vls.oz. 7.00			
" formate	⅛ oz.vls.oz. 7.00			
" hydrobromate	⅛ oz.vls.oz. 7.00			
" hydrochlorate, cryst.,—*U. S. Ph*.....	⅛ oz.vls.oz. 2.90			
" " powder,—Ph. Brit.............	⅛ oz.vls.oz. 2.90			
" hydrocyanate........................				
" hydro-iodate (hydriodate)............	⅛ oz.vls.oz. 8.00			
" hypo-phosphite	⅛ oz.vls.oz. 5.50			
" lactate	⅛ oz.vls.oz. 4.00			
" meconate (bi-meconate).............	⅛ oz.vls.oz. 3.75			
" nitrate	⅛ oz.vls.oz. 7.00			
" oleate, solution [20% Morphine]	⅛ oz.vls.oz. 3.00			
" phosphate..........................	⅛ oz.vls.oz. 7.00			
" phtalate,—*soluble in 4 parts Water*.—(The solution is very stable, and its subcutaneous administration is reported to be painless.)......................	⅛ oz.vls.oz. 4.00			
" saccharinate (*not* saccharate!)............ } *True salts of Morphine and Saccharin—which latter see!*				
" " bi-.................				
" salicylate...........................	⅛ oz.vls.oz. 5.50			
" sulphate, cryst.,—*U. S. Ph.*,—*but soluble* (as conforming to Ph. G. II) *in 14½ parts of Water!*	⅛ oz.vls.oz. 2.90			
" " with Strychnine	⅛ oz.vls.oz. 3.50			
" tannate	⅛ oz.vls.oz. 3.50			
" tartrate	⅛ oz.vls.oz. 3.85			
" valerianate	⅛ oz.vls.oz. 4.75			
Morphine and Codeine, hydrochlorate, see Salt, Gregory's.................				
" and Iron Oxide, hydrocyanate, see Morphine, ferro-hydrocyanate.......				
Morrhuol—(Regarded as the active principle of Cod-liver Oil.).....................	15 gr. .28			
Mountain-blue, artificial, see Copper, carbonate, blue..................				
" -green, artificial, see do., do., green ..				
Mucin,—from bile	15 gr. 1.00			
Muira puama,—(the new Aphrodisiac),—see under Fluid Extracts.................				

☞ When ordering, specify : "MERCK'S"!

	Containers incl.			
Mummy, true Egyptian	lb. 5.00			
Muraxid (Purpurate of Ammonium), dry	⅛ oz.vls.oz. 4.00			
Muscarine, nitrate	15 gr. 5.00			
" sulphate	15 gr. 5.00			
Musk-bags, empty				
Myrtol	oz. 1.75			

☞ When ordering, specify: "MERCK'S"!

	Containers incl.			
Nacarat, see Carmine, pure, in lumps				
Napelline.—Alkaloid from Aconitum napellus or from Aconitum lycoctonum	15 gr. 5.00			
Naphtha, Coal-tar-, see Benzene, anthracic				
" **Petroleum-,** see Benzin, petroleic				
" **vitriolic,**—so-called,—see Ether, sulphuric				
" **Wood-,** see Alcohol, methylic				
Naphthalene (Naphthalin), crude	lb. .20			
" perf. white, cryst.	lb. .25			
" " " resublimed	lb. .25			
" chem. pure, purified by Alcohol,—for internal use and antiseptic bandages				
cryst.	oz. .25			
powd.	oz. .25			
" Alpha-di-chlorated, (Alpha-Di-chlor-naphthalene), cryst.,—melting-point 35° C [95 F]	oz. 1.50			
" Alpha-mono-bromated, (Alpha-Mono-brom-naphthalene)	oz. 1.25			
" tetra-chloride	oz. 1.00			
Naphthalene Tapers	lb. 1.00			
Naphthalol, see **Betol**				
Naphtho-quinone (-chinone, -kinone), Alpha-	15 gr. 6.00			
Naphthol, Alpha-, recryst., perf. white.—(Recently brought to notice as a very efficient bactericide.)	oz. .40			
" **Beta-,** (Iso-Naphthol), purified	lb. .75			
" " " white, cryst.	oz. .25			
" " " " recrystallized	oz. .40			
" " " resublimed,—*medicinal!*	oz. .60			
Naphthol, Beta-, salicylate, see **Betol**				
Naphthol Tapers				
Naphtho-salol, see **Betol**				
Naphthyl-amine, crude	lb. 1.00			
" pure, white	oz. .40			
" chloride	oz. .45			
Narceine, pure	$\frac{1}{8}$ oz. vls. oz. 7.50			
" **acetate**	$\frac{1}{8}$ oz. vls. oz. 7.50			
" **hydrobromate**	$\frac{1}{8}$ oz. vls. oz. 7.50			
" **hydrochlorate, Merck, chem. pure.**—Prismatic crystals, easily soluble in Alcoholized Water; chemically neutral salt,—answering absolutely to the formula: $C_{23}H_{29}N\,O_9.H\,Cl$.—(A valuable sedative and hypnotic, preferred to Morphine,—especially in mental affections.)	$\frac{1}{8}$ oz. vls. oz. 8.50			
" **nitrate**	$\frac{1}{8}$ oz. vls. oz. 7.50			
" **sulphate**	$\frac{1}{8}$ oz. vls. oz. 7.50			
" **valerianate**	$\frac{1}{8}$ oz. vls. oz. 7.50			
Narcotine, pure	$\frac{1}{8}$ oz. vls. oz. 1.50			
" hydrochlorate	$\frac{1}{8}$ oz. vls. oz. 1.50			
" sulphate	$\frac{1}{8}$ oz. vls. oz. 1.50			
Natrio- (Sodio-) **Ethyl,** see Sodium, ethylate, etc., etc.				
Natrium, Natrum (Natron), and compounds,—see Sodium, etc.; and, Soda, etc.				
Neriin.—Glucoside from Nerium Oleander L.—(*Digitalein*-action claimed by Schmiedeberg.)				
Neurine, solution [25%]	15 gr. .45			
Nickel (Niccolum), double salts of, see "Nickel and —" (below!)				
" metallic, chem. pure	oz. 2.00			
" " —[98—99%], granulated	lb. 1.50			
" " —[98—99%], in cubes	lb. 1.50			
" " sheet and wire	lb. 2.00			

	Containers incl.
Nickel—(as above!),-metallic:-Anodes, cast or forged	lb. 2.00

Sizes of the Anodes in Millimetres:
(Extra sizes to order.)

a: forged.	b: cast.
300×200×2	100×100×3
300×200×1	150× 80×4
200×100×2	200×100×5
200×100×1	

" acetate	oz.	.50
" benzoate	oz.	.75
" bromide	oz.	.37
" carbonate	oz.	.25
" chloride	oz.	.20
" citrate	oz.	.50
" cyanide	oz.	1.50
" hydroxide, Niccolous, see Nickel, oxydulate, hydrated		
" iodide	oz.	1.00
" nitrate, pure	oz.	.25
" oxalate	oz.	.45
" oxide, black, (sesqui-oxide)	oz.	.25
" " chem. pure	oz.	.80
" " green, commercial	oz.	.25
" oxydulate (prot-oxide), hydrated, [Niccolous Hydroxide]	oz.	.75
" phosphate	oz.	.45
" sulphate	lb.	.60
" tartrate	oz.	.35
Nickel and Ammonium, chloride	oz.	.25
" " " citrate	oz.	.35
" " " nitrate	oz.	.35
" " " sulphate	lb.	.60
" and Potassium, sulphate	oz.	.35
Nicotine	⅛ oz.vls.oz.	4.00
Nigrosine,—Water-soluble ; and, Alcohol-soluble,—see under Aniline and Phenol Dyes : Black		
Nihil album, see Zinc, oxide, by dry process		
Niobe Essence, so-called, see Methyl, benzoate		
Niobium, metallic, pure	15 gr.	5.00
Nitre, cubic, see Sodium, nitrate		
" lunar, see Silver, nitrate, cryst.		
" prismatic, see Potassium, nitrate, chem. pure, cryst.		
" tabulated, see Potassium, nitrate, in flat drops		
Nitric Turpeth, see Mercury, nitrate, Mercurous, basic		
Nitro-aniline, meta-	oz.	3.00
Nitro-benz-aldehyd, ortho-		
Nitro-benzene (Nitro-benzol, Nitro-benzide) [so-called "Oil of Mirbane," "Essence of Mirbane"]—(erroneously called: "Artificial Volatile Oil of Bitter Almonds",—which latter see, under : Benz-aldehyd !);— light-colored	lb.	.60
Nitro-glycerine Tablets, Martindale's,—containing 0.00065 gramme [0.01 grain] of Nitro-glycerine each,—in boxes of 48 or 96 tablets		
Nitro-phenol, ortho-, colorless crystals,— melting-point 115°C [239 F]	oz.	1.00
" para-, yellow,—melt.-pt. 45°C [113 F].	oz.	1.75
Normal Solutions, titrated, (Test-solutions), see at End of List.		

When ordering, specify : "MERCK'S"!

	Containers incl.			

Oil—so-called—of Vitriol; free from Arsenic;—see Acid, sulphuric, crude ....
Oils, divers, (Olea varia)—[See, also: *Essential Oils*,—after: "Oils, divers"!]:—
   Almond; expressed,—true .... lb. .60
   "   "   " recent, English . lb. .75
   Amber; crude .... lb. .35
   anthelmintic, Chabert's, see Oil, Chabert's etc. ....
   Asphaltum .... oz. .50
   Beech, Europ.: fruit, (Beech-nut); expressed lb. 2.00
   Birch: wood; empyreumatic (*crude*),-[Birch Tar; Degutt, Daggett] .... lb. .40
   " do.;—*distilled* from the above,—see Essential Oils: Birch ....
   Cacao, see Butter, Cacao- ....
   Cade,—(Juniper Tar; Empyreumatic Oil of Juniper-wood) .... lb. .50
   Chabert's anthelmintic ....
   Chaulmoogra (*Chaulmugra*) [Gynocardia].. lb. 3.50
   Croton: seed; expressed,—(Oleum Tiglii).. lb. 1.50
   Egg-yelk (-yolk), [Oleum ovi]; recent ..... oz. .30
   empyreumatic, of Birch, see Oil, Birch....
   " of Juniper-wood, see Oil, Cade......
   " of Lignite, see Oil, Lignite ....
   " of Tobacco, see Oil, Tobacco ....
   Ergot; fatty,—expressed .... oz. .12
   ethereal, so-called; heavy,—(Heavy Oil of Wine),—see Oil, Wine, heavy..........
   —so-called—of Fern, ("Liquid Extract" or Oleoresin of Male Fern [Aspidium]), see Extracts: Male Fern,—ethereal.........
   —so-called—of Fusel-, see Alcohol, amylic, primary ....
   Haarlem, see Oil, sulphurated Linseed-, terebinthinated ....

☞ When ordering, specify : "**MERCK'S**"!

	Containers incl.		
Oils, divers;—*continued:*			
Henbane-leaves (Hyoscyamus); by digestion,—[*Oleo-infusion* of Hyoscyamus,—*Oleum coctum (infusum)* Hyoscyami*foliorum*]	lb. .60		
Henbane-seed; expressed, fatty	lb. .60		
Juniper-wood; empyreumatic,—see Oil, Cade			
Lignite; empyreumatic, — (Pyro-carbonic Oil), [Lignite Tar]	lb. 1.00		
Mace, so-called, see Oil, Nutmeg; expressed			
—so-called—of Male Fern, (Oleoresin of Aspidium), see Extracts: Male Fern,-ethereal			
—so-called—of Mirbane, (so-called "Essence of Mirbane"), see Nitro-benzene	·		
Nutmeg; expressed,—(Nutmeg-butter), [so-called "Oil of Mace"]	oz. .40		
Peach-kernel; fatty			
Persecot (Persico);—for preparing liquors	oz. 1.50		
Philosophers',—(Oleum Philosophorum)	lb. .50		
pyro-carbonic, see Oil, Lignite			
sulphurated Linseed-(Flaxseed-),—[Oleum Lini sulphuratum], (Balsam of Sulphur)	lb. .60		
do. do., terebinthinated; (Haarlem Oil; Dutch Drops), [Oleum Lini terebinthinatum sulphuratum], (Terebinthinated Balsam of Sulphur)	lb. .60		
Theobroma, see Butter, Cacao-			
Tobacco; empyreumatic,—*U. S. Ph.* 1870	oz. 2.00		
Wax; rectified, clear	oz. .50		
" " dark	oz. .40		
Wine; heavy,—(so-called "Heavy Ethereal Oil"), [Oleum Vini (æthereum) ponderosum].—(*Oleum æthereum, U. S. Ph.*, is a 50% [by volume] solution of this Oil, in Stronger Ether.)	lb. 5.00		
Wood-, — so-called,—("East-Indian Wood-oil," or : "East-India Copaiva Balsam," —so-called);—see Balsams : Gurjun			
N. B.—*See, also:—Oils,—so-called,—*flavoring : (Apple-; Fusel-; Grape-[Cognac-]; Pear-; Rum-),—*after:* "Essential Oils."			
Oils, Essential, see immediately below:—			
Essential Oils,—(inserted in alphabetical place of: *Oils, Essential*),—[Olea ætherea, volatilia, destillata], (Volatile Oils, Ethereal Oils, Distilled Oils):—			
Abies, see Essential Oil, Norway Pine			
Absinthium, see Ess. Oil, Wormwood			
Achillea (A. millefolium), see E. Oil, Yarrow			
Almond, Bitter, see Ess. Oil, Bitter Almond			
Amber .... rectified	lb. .60		
Angelica, European: root 30fold	oz. 8.00		
animal,—Dippel's .... twice rectified	oz. .40		
Anise: fruit, (Aniseed) .... duplex	lb. 5.50		
" Star-, (Chinese Anise), [Illicium]: fruit, [Badiane] .... duplex	lb. 4.50		
Arnica: flowers; true	½oz.vls.oz.30.00		
Artemisia maritima: flower-buds,—see Ess. Oil, Levant Wormseed			
Badiane, see Essential Oil, Anise, Star-			
Balm (Lemon-balm) [Melissa], German: herb	oz. 1.25		
Balsam Copaiva, see Essential Oil, Copaiva			
Bergamot: fruit-rind	lb. 3.25		
" do. .... sesquiduplex	lb.12.00	·	
Birch; distilled from Empyreumatic Birch-oil,—(*which compare*, under: Oils, divers)	lb. 1.50		

☞ **When ordering, specify : "MERCK'S" !**

	Containers incl.			
Essential Oils, — (inserted in alphabetical place of: *Oils, Essential*), — *continued:*				
Bitter Almond, — true.....................	lb. 5.00			
" " — artificial, — free from Hydrocyanic Acid; — (*not* = Nitrobenzene!); — see Benz-aldehyd				
Calamus (Sweet Flag): root (rhizome)......	lb. 3.25			
" do........................duplex	oz. 1.25			
Caraway-seed; from Dutch seeds..........	lb. 3.00			
" ....extra strong, — (*Carvol*)	oz. .50			
" ....sesquiduplex	oz. .75			
Cassia-bark, see Ess. Oil, Cinnamon, Chinese................................				
Cedar, Red, (Juniperus virginiana), see Ess. Oil, Red Cedar.........................				
Chamomile-flowers, German; blue, true...	oz. 3.00			
" Roman (English)...................	oz. 1.50			
Cherry-laurel: leaves.....................	oz. .75			
Cina, see Essential Oil, Levant Wormseed..				
Cinnamon, Chinese, (Cassia Cinnamon — Cassia lignea): bark..............duplex	oz. 1.25			
Cloves.................................	lb. 3.00			
" ....duplex	oz. .65			
— so-called — of Cognac, see Ether, oenanthic				
Copaiva (Copaiva-balsam).................	lb. 2.00			
Coriander: fruit ....sextuple	oz. 2.00			
Cubeb: fruit	oz. 1.25			
Cumin: fruit...................quadruple	oz. 1.50			
Eucalyptus; Australian, — from Eucalyptus *amygdalina*, (Peppermint-tree), and various allied species.....................	lb. 2.00			
Eucalyptus *globulus:* leaves; dextrogyrate N. B. — *See, also*: Eucalyptol; and, Eucalyptol, chem. pure!	lb. 2.50			
Fennel: fruit............................	lb. 2.00			
" " ....duplex	lb. 4.00			
Gaultheria, see Essential Oil, Wintergreen .				
Ginger: root (rhizome); true..............	oz. .75			
Grape-marc (Vitis vinifera), — so-called, — see Ether, oenanthic				
Hops...................................	oz. 4.50			
Illicium (Star-anise), see Essent. Oil, Anise, Star-..............................				
Juniper (Juniperus communis): berries; best	lb. 2.50			
" do.; do......................20fold	oz. 2.00			
Juniper (Juniperus communis): wood	lb. .60			
Juniperus virginiana, see Ess. Oil, Red Cedar				
Laurel (Sweet Bay): fruit.................	oz. 1.00			
Lavender: flowers ....sesquiduplex	oz. 1.00			
Lemon: fruit-rind	lb. 2.25			
" " 30fold	oz. 5.00			
Lemon-balm, see Ess. Oil, Balm..........				
Levant Wormseed, (Cina; Santonica; — Semen contra; Semen sanctus): [the flower-buds of Artemisia maritima]............	oz. .25			
Matico: leaves...........................	oz. 4.00			
Melissa, German, see Ess. Oil, Balm.......				
Milfoil (Millefolium), see Ess. Oil, Yarrow..				
Mint, Curled, (Mentha crispa): herb, -double				
Mint, Pepper-, see Ess. Oil, Peppermint...				
" " Chinese or Japanese, -(Poho-oil), see Ess. Oil, Peppermint, Chinese; *true*....................				
Mustard, Black: seed; true...............	lb. 12.00			
" " — artificial, — (Allyl Sulphocyanate [Thio-cyanate], — synthetically prepared)....	lb. 7.00			
Norway Pine, (Norway Spruce Fir), [Abies]: shoots..............................	lb. 1.75			
Orange: fruit-rind.................30fold	oz. 6.00			

☞ **When ordering, specify: "MERCK'S"!**

	Containers incl.		
Essential Oils, — (inserted in alphabetical place of: *Oils, Essential*), — *continued:*			
Pepper, Black...........................	oz. .50		
Peppermint: herb ....double	oz. 1.50		
Peppermint, Chinese (Japanese); *true,* — [Poho-oil];—only in original flasks......			
Pine-needles (Leaves of Pinus sylvestris)...	lb. 1.50		
Pine-shoots, — (Oleum templinum), — see Ess. Oil, Pinus pumilio...............			
Pinus pumilio, (Hungarian Balsam tree): shoots;—[Oleum templinum]......	oz. .75		
" sylvestris, see Ess. Oil, Pine-needles..			
Poho-, see Ess. Oil, Peppermint, Chinese ..			
Red Cedar, (Juniperus virginiana): root ...	lb. 1.00		
Santal, East-Indian: wood, (Sandal-wood), [Yellow Saunders, White Saunders]	lb. 7.00		
" West-Indian: wood................	lb. 4.00		
Santonica (Cina), see Ess. Oil, Levant Wormseed...............................			
Sassafras: wood; true	lb. 1.00		
" " " ....double	lb. 4.00		
Savin: tops	lb. 1.25		
Semen cinæ, (Semen *contra; S. sanctum; S. santonici*), see Ess. Oil, Levant Wormseed			
Spiræa ulmaria, (Meadow-sweet), see Acid, salicylous............................			
Star-Anise, see Ess. Oil, Anise, Star-			
Sweet Flag, see Ess. Oil, Calamus			
Tansy: leaves	lb. 12.00		
Templin (Pine - shoot), see Ess. Oil, Pinus pumilio...............................			
Thyme: herb...................quintuple	oz. 1.00		
Turpentine	lb. .40		
" ....rectified	lb. .50		
" —Hydrochlorates of, — see Turpentine-oil, mono-hydrochlorate; and, do., di-hydrochlorate			
Valerian: root..........................	oz. .75		
Vitis vinifera, (Grape-marc),—so-called,— see Ether, oenanthic.................			
Wintergreen (Gaultheria): leaves..rectified	oz. .50		
Wormseed, Levant-, (Santonica), see Ess. Oil, Levant Wormseed................			
Wormwood (Absinthium): herb; true......	lb. 8.00		
" do.; do.....................10fold	oz. 2.50		
Yarrow (Milfoil — Achillea millefolium): flowering herb........................	oz. 1.50		
Oils—*so-called,*—**flavoring:**			
Apple-, see Amyl, valerianate............			
Cognac- (*Grape*-), see Ether, oenanthic....			
Fusel-, see Alcohol, amylic...............			
Pear-, see Amyl, acetate			
Rum-, see Essential Spirits: Rum,—concentrated............................			
Ointment, blue, (Unguentum Hydrargyri cinereum; Ph. G. II),—[33⅓% Mercury]	lb. .60		
" do., duplex,—[50% Mercury]	lb. .80		
" " with Cerato of Nutmeg-butter, (cum Cerato Myristicæ; cum Balsamo Nucistæ),—[50% Mercury]			
" " with Lanolin,—[50% Mercury]...	lb. 2.00		
" Chaulmoogra (Gynocardia),—[1 part of Chaulmoogra-oil: 3 of Vaselin]......	lb. 2.00		
Ointments on Mollin (a new Ointment-base), see **Mollin Ointments**...................			
Oleandrin.—Glucoside from Oleander (Nerium, O., Linné).—[*Digitalin*-action claimed by Schmiedeberg.]			
Olea ætherea (*volatilia, destillata*), see Oils, Essential, [*Essential Oils*]			

☞ When ordering, specify: **"MERCK'S"!**

MERCK'S INDEX. 103

	Containers incl.			
Olea cocta (*infusa*), see *Oleum coctum* etc....				
Olea varia, see Oils, divers				
Olein, see Acid, oleic......................				
Oleo-infusion (*Oleol*) [Olcum coctum (*infusum*)] of **Henbane-leaves** (Hyoscyamus), see under Oils, divers....................				
Oleoresins :				
Aspidium (Male Fern), *U. S. Ph.*, see Extracts: Male Fern,—ethereal				
" —Ph. G. II,—see *same*,—free fr. Ether				
Capsicum,— *U. S. Ph.*,—(Ethereal Extract of Guinea Pepper—of Capsicum fastigiatum).............................	oz. .75			
Other Ethereal Extracts, (Oleoresins):—				
Brayera (Kousso)..........				
Cantharides................				
Cubeb..................... See likewise				
Eucalyptus				
Indian Hemp (Cannabis)... under				
Kamala (Rottlera).........				
Matico .... *Extracts.*				
Mezereon..................				
Phellandrium..............				
Valerian....				
Oleum æthereum ponderosum, so-called, (*Oleum Vini ponderosum*), see Oils, divers: Wine; heavy.—[*Oleum æthereum, U. S. Ph.*,—see *remark* after same.]				
Oleum coctum (*infusum*) **Hyoscyami foliorum**, see Oils, divers: Henbane-leaves				
Ononin.—Glucoside from the root of Ononis spinosa—Rest-harrow....................	15 gr. 1.00			
Ophioxyline. — Alkaloid from Ophioxylon serpentinum,—acc. to Prof. Bettink				
Ophthalmic Stone, so-called, see Copper, aluminated...............................				
Opianyl, see Meconin				
Orcin (Symmetric meta-Di-oxy-toluene).— From lichens of the Rocella and Lecanora families	oz. 2.00			
Orellin, r'd, see Bixin				
Ormosine, cryst.—Alkaloid from the seed of Ormosia dasycarpa................				
" hydrochlorate, cryst....................	15 gr. 3.00			
Orpiment, see Arsenic, Yellow sulphide....				
ortho-Amido-phenol, hydrochlorate, see Amido-phenol, ortho-, etc.................				
ortho-Nitro-benz-aldehyd, see Nitrobenz-aldehyd, ortho-				
Osmium, metallic........................	15 gr. 3.00			
" tetr-oxide, see Acid, per-osmic, anhydr.				
Osmium-Iridium alloy, (*Osm-iridium; Irid-osmium*)..................................	15 gr. 1.50			
Ostrich Pepsin, see Pepsin, Ostrich.......				
Ouabain—[$C_{30}H_{46}O_{12}$].—Crystallized Glucoside from the Ouabaio-tree—(an aqueous extract from whose root and bark forms the arrow-poison of the East-African Comalis).— [A heart-poison hypodermically.]..........				
Ox-amide...............................	oz. 2.00			
Ox-aniline, ortho-, hydrochlorate, see Amido-phenol, ortho-, etc.				
Ox Gall, inspissated, *U. S. Ph.*, (also called: *Extract of* Ox Gall), see Gall, Ox-.				
" " purified, dry, see Sodium, cholcate.				
Oxide, magnetic, see Iron, oxide, black...				
Oxy-acanthine, pure...................	15 gr. 1.50			
" hydrochlorate	15 gr. 1.50			
Oxy-benz-aldehyd, ortho-, see Acid, salicylous..				

☞ **When ordering, specify : "MERCK'S"!**

	Containers incl.			
Oxygen Hydrate... ⎫ **Oxygenated Water,** ⎬ see **Hydrogen Per-** so-called .... ⎭ oxide, etc.; etc. ...				
Oxy-phenyl-benzyl-ketone (-acetone), see **Benzoin Crystals**.....................				
Oxy-toluol-tropine, etc., see **Homatropine Merck-Ladenburg,** etc.				

☞ When ordering, specify: "**MERCK'S**"!

	Containers incl.			
Palladium, metallic,—sheets or wire	15 gr. 2.00			
" do., — black precipitate, (Palladium Black [Mohr])	15 gr. 2.00			
" chloride, dry	15 gr. 2.00			
" " solution	⅛ oz. vls. oz. 8.00			
" nitrate, dry	15 gr. 2.00			
" " solution	⅛ oz. vls. oz. 8.00			
Palladium and Sodium, chloride, dry	15 gr. 1.00			
Palladium Black, ⎱ see Palladium, metallic,				
" **Mohr** ⎰ —black precipitate .				
Pancreatin, pure, absolute	oz. .75			
" " active	oz. .45			
" " in scales	oz. .85			
" " — solution in Glycerin, [1:10],-(Glycerolate [Glycerite] of Pancreatin	lb. 2.00			
N.B.—*Compare, also:—* Solutions: pancreatic.				
" saccharated	oz. .50			
" with Starch	oz. .35			
N.B. —*See, also:* TRYPSIN (the Albumen-solving constituent of Pancreatin)!				
Pancreatin-Pepsin	oz. .45			
Papaverine Merck:				
pure	⅛ oz. vls. oz. 6.00			
hydrochlorate	⅛ oz. vls. oz. 6.00			
nitrate	⅛ oz. vls. oz. 6.00			
phosphate	⅛ oz. vls. oz. 6.00			
sulphate	⅛ oz. vls. oz. 6.00			
Papaw Juice, (Succus Caricæ Papayæ), see **Juice of Papaw**				
Papayotin Merck,—from Papaw Juice;—peptonizes 200 parts of freshly expressed Blood-fibrin.—(Used with especial success as a solvent of diphtheritic membranes.)	15 gr. .50			
Paper, Congo-, (Prof. Riegel's "Gastric" Testpaper), see **Congo Paper**				
" Wax-	quire .30			
" Litmus-, red or blue, (red or blue Testpaper)	quire .75			
" Turmeric- (*Curcuma-*), [yellow Testpaper]	quire 1.00			
Papers, medicated,—for Ophthalmology,— see under Atropine and Physostigmine				
para-Acet-phenetidin, see **Phen-acetin**				
para-Cotoin, see **Cotoin, para-**				
Paraffin, solid, — solidifying-point 46–48° C [114.8-118.4 F]	lb. .20			
" do.,—solidif.-pt. 52–53°C [125.6-127.4 F]	lb. .25			
" " " 56–58°C [132.8-136.4 F]	lb. .30			
" " —Ph. G. II,—melting-point 74–76°C [165.2-168.8 F]	lb. .50			
" liquid,—Ph. G. II	lb. .60			
para-Globulin, see **Globulin, para-**				
Paraguay roux, see Tinctures: Spilanthes; compound				
Par-aldehyd Merck, chem. pure, (*absolutely pure*), —of unexceptionable quality	lb. 2.50			
Parillin (Pariglin, Sarsaparin), see **Smilacin**				
Parsley-camphor, see **Apiol, solid, cryst., white**				
Pear-oil, so-called, see **Amyl, acetate**				
Pearl-ash, see **Potassium, carbonate**				
Pelletierine (Punicine) **preparations:**				
Pelletierine, medicinal,—(Pelletierine and Iso-pelletierine),—pure	15 gr. 2.50			
" sulphate, pure	15 gr. 1.75			
" " " —10%-solut.				
" tannate	15 gr. .75			
" valerianate	15 gr. 2.50			

☞ **When ordering, specify : "MERCK'S"!**

	Containers incl.		
Pelletierine (Punicine) preparations,—*continued:*			
Methyl-pelletierine, pure,—oily liquid	15 gr. 3.00		
Pseudo-pelletierine, pure, crystallized	15 gr. 2.50		
" hydrochlorate, white, cryst.	15 gr. 2.00		
" sulphate, white, cryst.	15 gr. 1.75		
Pentane (Amyl Hydride), crude, see Eupione			
Scale or Pwd. { **Pepsin Merck 1 : 1000,**—*digests* 1000 *times its weight* } *of coagulated Albumen.* (My Pepsins are powerfully active preparations of uniform strength and high quality.)	oz. .75		
Pepsin Merck 1 : 1500,—*digests* 1500 *times its weight*	oz. 1.00		
Pepsin Merck 1 : 2000,—*digests* 2000 *times its weight*	oz. 1.25		
☞ *All other strengths to order!*			
Pepsin, pure, soluble, in scales } — *Strength corresponding to Ph. G. II.*	oz. .60		
" " **granulated**	oz. .50		
" " clearly soluble, powder,—Ph. G. II. . .	oz. .40		
" pure,-solution in Glycerin,-*concentrated*	oz. .30		
" hydrochlorate, clearly soluble,—powder	oz. .50		
" " clearly soluble,—extract form . . .	oz. .50		
" with Dextrin,—yellow	lb. 2.00		
" " Starch,—white	lb. 2.00		
Pepsin, Ostrich-	oz. .75		
Pepsin Essence,—acc. to Dr. Liebreich,— in original bottles	bottle 1.00		
Pepsin, Lacto-, (also called "Lactated Pepsin"),—[sometimes mis-called "Lacto-peptine"]	oz. .50		
" **Pancreatin-,** see Pancreatin-Pepsin . .			
" **Peptone-,** etc., see Peptone-Pepsin, etc.			
" **Ptyalin-,** see Ptyalin-Pepsin			
Pepsin Wine, (Vinum pepsini,—Ph. G. II)	lb. 1.25		
Peptone, soft, from Meat, } Pure Meat Peptones,	lb. 2.00		
" dry, " " } free from Par-albumin	lb. 3.00		
[*The above Dry* Meat Peptone answers to 7-8 times its weight of fresh meat.]			
" dry, from Albumen	oz. .50		
Peptone, bismuthated, see **Bismuth, peptonized** . .			
Peptone-Pepsin, phosphate	oz. .40		
" tartrate	oz. .35		
Peptone-Quinine, see **Quinine,** peptonized			
Pereirine, pure	15 gr. 3.00		
" hydrochlorate	15 gr. 2.50		
Petroleum Benzin . . . . } see Benzin,			
" **Naphtha** . . . . } petroleic			
" **Ether,** see Benzin, petroleic, boil.-pt. 50-60° C,—(*Benzinum, U. S. Ph.*)			
Peucedanin (Imperatorin)	15 gr. .60		
Phen-acetin (para-Acet-phenetidin).—Colorless, inodorous, insipid crystals, — readily soluble in Alcohol, less so in Water; melting-pt. 132.5° C [270.5 F].—(A new antipyretic.)	oz. 1.25		
Phen-acetolin	⅛ oz. vls. oz. 4.00		
Phen-anthrene	oz. .50		
Phenetol (Ethyl Phenate [Carbolate]; Ethylic Ether of Carbolic Acid; Ethylo-phenic [Ethylo-carbolic, Phenol-ethylic] Ether) —[also called: Salithol]			
Phen-oxy-Caffeine, see **Phenyl-oxy-Caffeine**			
Pheno-Resorcin (-Resorcinol)	oz. .50		
Phenol (so-called "Phenyl Hydrate"), see **Acid, carbolic**			
" camphorated, (*Phenol-Camphor*), see **Camphor, phenolated**			
" iodized, see **Acid, carbolic, iodized**			
" salicylate, see **Salol**			
Phenol-Cocaine, see **Cocaine phenate**			
Phenol Dyes (*Colors*), see under **Aniline and Phenol Dyes**			

☞ When ordering, specify: "**MERCK'S**"!

---

* *See, also, page* 167 !

	Containers incl.		
Phenol-Glycerin, see Acid, carbolic,—solution in Glycerin..................			
Phenol-phtalein, pure,—Ph. G. II.............	oz. 1.50		
Phenol-Quinine, see Quinine, phenate....			
Phenyl, bromide, see Mono-brom-benzene..			
" chloride, see Mono-chlor-benzene......			
" hydrate,—so-called,—see Acid, carbolic.			
" hydride,—so-called, —see Benzene, anthracic, chem. pure, crystallizable...			
" iodide, see Mono-iod-benzene.....			
Phenyl-acet-amide,—medicinal,—see **Antifebrin**..			
" mono-bromated, see Brom-phenyl-acet-amide, mono-...................			
Phenyl-amine, see Aniline................			
Phenyl-glucos-azone....................	15 gr. .60		
Phenyl-hydrazine, pure.................	oz. 1.25		
" hydrochlorate......................	oz. 1.00		
Phenyl-lactos-azone......................	15 gr. .75		
Phenyl-methane, see Toluene............			
Phenyl-methyl-ketone, (-acetone), see **Hypnone**...			
Phenyl-oxy-Caffeine (Phen-oxy-Caffeine)..	15 gr. .75		
Phenylene-di-amine, meta-, hydrochlorate, see Di-amido-benzene, meta-, hydrochlorate.................			
Philosophers' Wool, so-called, see Zinc, oxide, by dry process..................			
Phloretin (Phloretic Acid), cryst.—Fractional derivative of Phlorizin............	15 gr. .60		
Phlorizin (Phloridzin, Phlorrhizin).—Glucoside from the root-bark of the Apple-tree	½ oz.vls.oz. 3.00		
Phloro-glucin (-glucol, -glucinol), chem. pure.—free from Di-resorcin;—melting-point 210° C [410 F]............................	15 gr. .25		
Phosphine, so-called, see Aniline and Phenol Dyes: Yellow, Chrys-aniline..........			
Phosphorus, amorphous (red).............	lb. 2.25		
" vitreous (yellow), [also called "*Crystallized* Phosphorus"],—*Phosphorus, U. S. Ph.*...........................	lb. 1.10		
" bromide............................			
" iodide.............................	oz. 1.50		
" oxy-chloride......................	oz. .50		
" penta-bromide...................	oz. .60		
" penta-chloride [P Cl$_5$]...............	oz. .50		
" pent-oxide [P$_2$O$_5$], see Acid, phosphoric, anhydrous......................			
" tri-chloride [PCl$_3$]...................	oz. .50		
" tri-sulphide, (Thio-phosphorous Anhydride), [P$_2$S$_3$],-melt.-pt. 290°C [554°F]	oz. .75		
Physostigmine (Eserine), chem. pure, cryst. —Alkaloid from Calabar Bean..	grain .25		
" (Eserine), citrate.............	grain .20		
" " hydrobromate, cryst.........	grain .20		
" " hydrochlorate, cryst........	grain .20		
" " nitrate..................	grain .20		
" " salicylate, cryst., Merck,—*U.S. Ph.* and Ph. G. II......	grain .15		
" " sulphate, white, Merck......	grain .15		
" " tartrate................	grain .20		
Physostigmine Discs, (*Eserine* Discs, Calabar Discs),—in tubes of 100.........			
" **Gelatin,** (*Eserine* Gelatin, Calabar Gelatin),—in sheets for 25 applications..			
" **Paper,** (*Eserine* Paper, Calabar Paper),—in books for 100 applications.....			
Physostigmine, Pseudo-, pure.—Alkaloid from *Nux Cali*, (Pseudo-Calabar Bean).....	grain 1.00		
Picoline, chem. pure....................	oz. 1.50		
Picro-podophyllin.......................	15 gr. .50		
Picro-toxin............................	½ oz.vls.oz. 5.00		

(All these preparations are perfectly pure.)

☞ **When ordering, specify: "MERCK'S"!**

MERCK'S INDEX.

	Containers incl.
Pilocarpidine Harnack-Merck. nitrate, cryst.	15 gr. 3.00
Pilocarpine. pure (All these preparations are perfectly pure, and free from Jaborine.)	grain .13
" hydrobromate	grain .13
" hydrochlorate. cryst., chem. pure, — Ph. G. II	grain .07
" nitrate, cryst.	grain .07
" salicylate	grain .10
" sulphate	grain .07
" tannate	grain .07
" valerianate	grain .15
Pink Salt, (Dyers' Salt), see Tin and Ammonium, chloride	
Piperidine	oz. 1.00
" hydrochlorate	oz. 1.50
Piperine, pure	oz. .79
Piperonal, chem. pure, (Methylene-proto-catechu-aldehyd)	15 gr. .50
" for perfumery,—also called **Heliotropin**.	15 gr. .50
Pix, etc., see Tar, etc.	
Plaster, adhesive, English, — spread, — in 6-yd. rolls	
" Ichthyol-, see under Ichthyol preparat.	
" Lead-, simple, (Diachylon-plaster; Litharge-plaster)	
" Mezereum-and-Cantharides-,—spread	
Platina, etc., see Platinum, etc.	
Platina Black (Mohr), see Platinum, metallic, black precipitate	
Platina Sponges, prepared and mounted for Hydrogen lamps.—(*See, also:* Platinum, metallic, spongious.)	doz. 1.80
Platinum (Platina), double and triple salts of, see (*below*):— "Platinum double Chlorides"; "Platinum double Cyanides"; "Platinum triple Cyanides"; "Platinum, divers double Salts";— *also:* "Platos-amine, di-, sulphate".	
" metallic, wire and sheets	15 gr. .50
" " spongious.—(*See, also:* Platina Sponges, for Hydrogen lamps.)	15 gr. .60
" " black precipitate, (Platina Mohr, Platinum Black)	15 gr. .60
" cyanide, Platinous, (Platinum Cyanuret)	15 gr. 1.00
" bi-chloride (*di*-chloride,—*formerly* called *proto*- or *mono*-chloride), [chloruret], (Platinous Chloride)	15 gr. 1.00
" iodide	15 gr. 1.00
" nitrate	15 gr. .75
" tetra-chloride (*per*-chloride,—*formerly* called *bi*- or *di*-chloride), [Platinic Chloride], dry	¼ oz.vls.oz. 6.00
" " —solution [1 : 20]	⅛ oz.vls.oz. 1.00
" " " [1 : 10]	⅛ oz.vls.oz. 1.50
Platinum double Chlorides:	
Platinum bi-chloride and Ammonium chloride, (Platin-ammonium Chloride), —[Pt Cl$_2$. 2 NH$_3$ Cl]	15 gr. 1.00
" tetra-chloride and Ammonium chloride, (Platinum Sal-ammoniac), dry, —[Pt Cl$_4$. 2 NH$_3$ Cl]	15 gr. .65
" do. do. do. do., cryst.	15 gr. 1.00
" and Ammonium, chloruret, (Ammonio-Platinous chloride), cryst.	15 gr. 1.25
" and Barium, chloride,—crystallized with 4 molecules of Water	15 gr. 1.00
" bi-chloride and Potassium sesquichloride, cryst.	15 gr. 1.25
" tetra-chloride and Potassium sesquichloride, dry	15 gr. .60
" do. do. do. do., cryst.	15 gr. 1.00

☞ When ordering, specify : "**MERCK'S**" !

	Containers incl.		
Platinum double Chlorides,—*continued:*			
Platinum and Sodium, chloride, cryst.	15 gr. 1.25		
" " " " dry	15 gr. .65		
" -tetr-amine and Platinum, bi-chloride, (Platoso-di-ammonium Chloro-platinite), [Magnus's "Green salt"],—(Pt [NH$_3$]$_4$Cl$_2$. Pt Cl$_2$)			
Platinum double Cyanides:			
Platinum and Ammonium, cyanide, cryst.	15 gr. 1.00		
" and Barium, cyanide, cryst.	15 gr. 1.25		
" and Calcium, " "	15 gr. 1.00		
" cyanuret and Copper cyanide, (Platino-cupric cyanide)	15 gr. 1.25		
" and Lead, cyanide, cryst.	15 gr. 1.25		
" and Magnesium, cyanide, cryst.	15 gr. 2.00		
" and Potassium, " "	15 gr. 1.25		
" " " sesqui-cyanide, cryst.	15 gr. 1.25		
" and Sodium, cyanide, cryst.	15 gr. 1.50		
" and Strontium, cyanide, cryst.,—with 5 molecules of Water	15 gr. 1.25		
" " " do., do.,—with 4 molecules of Water	15 gr. 1.25		
" and Yttrium, cyanide, large cryst.	15 gr. 2.50		
Platinum triple Cyanides:			
Platino - Ammonio - cyanuret and Cupric cyanide, (Platino-Ammonio-Cupric cyanide), cryst.	15 gr. 1.25		
" -Calcio-Ammonio-cyanuret, cryst.	15 gr. 1.25		
" -Potassio-Lithio- " "	15 gr. 2.00		
" -Potassio-Sodio- " "	15 gr. 1.50		
Platinum, divers double Salts:			
Platinum and Ammonium ... } sulpho-cyanate—(thio-cyanate; rhodanide),— cryst. {	15 gr. 1.00		
" and Barium	15 gr. 1.00		
" and Potassium	15 gr. 1.25		
" " do., bromide, cryst.	15 gr. 1.25		
" " " iodide, "	15 gr. 1.25		
" cyanuret, (Platinous cyanide), and Potassium Chloride	15 gr. 1.25		
Platinum Black, } see Platinum, metallic, black precipitate, {			
" **Mohr**			
" **Sal ammoniac,** see Platinum double Chlorides: Platinum tetra-chloride and Ammonium chloride: dry; and, cryst.			
Platinum Sponges, prepared and mounted for Hydrogen lamps, see Platina Sponges			
Platos-amine, di-, (Di-platos-amine), sulphate, cryst.	15 gr. 1.25		
Plumbago, see Graphite			
Plumbum, and compounds, see Lead, etc.			
Podophyllin, chem. pure } Both yield a perfectly clear solut. in Alcohol. {	oz. .60		
" **pure,—Ph. G. II**	oz. .40		
Podophyllo-toxin,—acc. to Podwyssotzki	15 gr. .30		
Polishing-powder (so-called "Putty-powder"), see Tin, oxide, grey			
Polygalin (Polygalic Acid), see Senegin			
Populin	15 gr. 1.50		
Potassa (Kali), caustic, chem. pure, Merck, see Potassium, hydroxide, chem. pure, Merck			
" do.,—*other grades and forms,*—see Potassium, hydroxide, etc., etc., etc.			
" *U. S. Ph.*, see Potassium, hydroxide, purified, in sticks			
Potassa, Anthraco-; and do., sulphurated;—see Anthraco-potassa; etc.			
Potassa, antimonio-sulphurated, *crude*, (Liver of Antimony), [so-called "Unwashed Brown Oxide of Antimony"],—(improperly called, also: "Antimonio-sulphide of Potassium").—[**Do.,** do., *washed,*—see *next page!*]	lb. .75		

☞ **When ordering, specify: "MERCK'S"!**

	Containers incl.
Potassa, antimonio-sulphurated, *washed* (lixiviated). — [Crocus (Saffron) of Antimony; Crocus metallorum], -- (so-called "Washed Brown Oxide of Antimony").	lb. 1.00
N.B.—*See, also:* Potassa, antimonio-sulphurated, *crude.*—(*preceding page!*).	
Potassa, cantharidated, see Potassium, cantharidate	
Potassa, sulphurated, (Liver of Sulphur; Potassic Liver of Sulphur), [improperly called "Potassium Sulphide"], —*crude;*—for baths	lb. .30
" do.,—*purified;*—from Purified Potassium Carbonate:— *Potassa sulphurata, U. S. Ph.*	lb. 1.00
" do.,-*pure,*-from Pure Potassium Carb.	lb. 1.25
Potassa with **Lime,** *U. S. Ph.,—(Potassa-Lime);-also:* Vienna Caustic Powder;-*and:* Filhos's Caustic:— see Potassium, hydroxide, with Lime: [1:1];—[2:1];—and, [4:1]	
Potassa Alum, see Alum, potassic	
Potassa Prussiates:	
Red, pure ....) see Potassium, ferrid-cy- " commercial, ∫ anide, etc.	
Yellow, chem. pure, ⎫ see Potassium, ferro- " commerc'l. ⎬ cyanide,—*U.S.Ph.,* " with Urea.. ⎭ etc.	
Potassio-Phtal-imide, see Potassium, imido-phtalate	
Potassium (Kalium), double and triple salts of, see "Potassium and —" (below!)	
" metallic	¼ oz.vls. oz. 2.25
" acetate, (Terra foliata tartari), purified, commercial	lb. .48
" " purified, white	lb. .75
" " " fused	lb. 1.50
" " pure,— *U. S. Ph.* and Ph. G. II	lb. .75
" " " fused	lb. 2.00
" " chem. pure	lb. 1.50
" aceto-wolframate (aceto-tungstate)	oz. .40
" æthylo-sulphate, see Potassium, ethylo-sulphate	
" antimonate, *pharmacopeial* (Ph. Bor. VI), —[Washed (purified) Diaphoretic Antimony], (so-called "White Oxide of Antimony, Ph. Bor. VI"; also called: *Calx Antimonii* [Stibii]):—[*principally:* K SbO₃]	lb. 1.00
" do., *do.*, in troches (lozenges)	lb. 1.50
" antimonate, *crude,*—(Unwashed Diaphoretic Antimony), [so-called "Unwashed Diaphoretic Oxide of Antimony"]	lb. .85
" antimonate, *pure* by assay	oz. .30
" antimonio-sulphide, — so-called, — *see* Potassa, antimonio-sulphurated, *crude*	
" arseniate (arsenate)	oz. .14
" " pure	oz. .20
" arsenite, crude	oz. .14
" " pure	oz. .20
N.B.—*Fowler's Solution,* see Solutions: Potassium arsenite, *U. S. Ph.*	
" benzoate	oz. .64
" bi-borate	oz. .20
" bi-carbonate (acid carbonate), pure, cryst.,—*U. S. Ph.* and Ph. G. II	lb. .28
" " chem. pure, cryst.	lb. .50
" bi-chromate, chem. pure, cryst., — *U. S. Ph.*	lb. .50
" " pure, fused	lb. 2.00

☞ **When ordering, specify : " MERCK'S " !**

	Containers incl.			
Potassium, bi-chromate, —(continued!),—				
commercial, cryst.................	lb. .25			
" do., do., fused......................	lb. .75			
" bi-fluoride............................	oz. .45			
" bin-oxalate, (Salt of Sorrel—Sal Acetosellæ), [so-called "Essential Salt of Lemons"]	lb. .40			
" " pure...........................	lb. .75			
" bi-phosphate	lb. 2.50			
" bi-sulphate, (Hydro-mono-potassic Sulphate)........................	lb. .50			
" " chem. pure, cryst.	lb. .75			
" " " " fused...............	lb. 1.00			
" " pure, cryst.	lb. .60			
" " " fused	lb. .75			
" bi-sulphite (acid sulphite), chem. pure, cryst., — abt. 87% of KHSO$_3$);—readily soluble in Water	lb. 2.00			
" bi-tartrate (acid tartrate), cryst., [Crystals of Tartar], (Purified Tartar)	lb. .75			
" " powder, (Powdered Crystals of Tartar), [Pure powdered Tartar]	lb. .80			
" " pure, powder, (Pure Cream of Tartar),—free from metals........	lb. .85			
" " chem. pure, powder) free from " " do. do., cryst.,—U. ⎬ metals and S. Ph. ....) from Lime, [—conforming to Ph. G. II	lb. .90 lb. .85			
" borate	oz. .18			
" bromate, pure,—Ph. G. II; —(perfectly pure: [100%]).....................	oz. 1.00			
" bromide, chem. pure, powder,—Ph. G. II	lb. 1.00			
" " " " cryst.,—U. S. Ph. and Ph. G. II..	lb. 1.00			
" " " " disturbed crystals, —Ph. G. II....	lb. 1.00			
" bromino-arsenite	oz. 1.50			
" " -salicylate	oz. 6.00			
" cantharidate, (Cantharidated Potassa)..	15 gr. 5.00			
" carbolate, see Potassium, phenate.....				
" carbonate, (Pearlash), [80–84% of pure]..	lb. .20			
" " [90–92% of pure]...............	lb. .25			
" " [95–98% " "]............∴.....	lb. .30			
" " twice purified	lb. .35			
" " pure,—U. S. Ph. and Ph. G. II,—from the Bi-tartrate.—(This grade of Potassium Carbonate is also called: Salt of Tartar;—not to be confounded with: "Essential Salt of Tartar"= Tartaric Acid!)...	lb. .60			
" " chem. pure...................	lb. .70			
" carbonate, acid, see Potassium, bi-carb.				
" **caustic oxide, chem. pure, Merck, etc., see Potassium, hydroxide, etc., etc.**.........				
" chlorate, cryst.	lb. .40			
" " powder	lb. .40			
" " pure, cryst.,—U. S. Ph. and Ph. G. II....................	lb. .50			
" " " powder,—Ph. G. II........	lb. .50			
" chloride, crude,—[about 98%].........	lb. .25			
" " chem. pure	lb. .50			
" chromate, yellow, chem. pure.........	lb. 1.25			
" " " purified.............	lb. .70			
" " " commercial	lb. .35			
" cinnamate,—from pure Cinnamic Acid; —very freely soluble in Water......	oz. 2.00			
" citrate, pure,—U. S. Ph..............	lb. 1.50			
" cobalti-cyanide, (Cobalto-tri-potassic Tri-cyanide), anhydrous,—readily soluble in Water				

☞ **When ordering, specify : "MERCK'S"!**

		Containers incl.			
Potassium, cyanate		oz. 1.50			
" cyanide, [about 30%], fused,] plates {		lb. .50			
" " [" 40%], "		lb. .55			
" " [" 45%], " or		lb. .60			
" " [" 50%], "		lb. .65			
" " [" 60%], " sticks		lb. .75			
" " pure, [about 85%],—in plates		lb. 1.25			
" " " " " —in sticks		lb. 1.30			
" " " [96 to 100%],—U. S. Ph.		lb. 2.00			
" " chem. pure		lb. 4.00			
" ethylo-sulphate (sulpho-vinate)		lb. 2.00			
" ethylo-thio-carbonate, see Potassium, xanthogenate					
" ferrid-cyanide (ferri-cyanide), [Red Prussiate of Potassa], (Potassio-*ferric* cyanide, so-called),—pure		lb. 1.50			
" " commercial		lb. 1.00			
" ferri-ferro-cyanide, (Soluble Prussian Blue), see Iron, cyanide, blue,—so-called,—*soluble*					
" ferro-cyanide,—(Yellow Prussiate of Potassa), [Potassio-*ferrous* cyanide, so-called],—chem. pure,—U. S. Ph.		lb. 1.00			
" " commercial		lb. .60			
" " with Urea		lb. 4.00			
" fluoride		lb. 2.00			
" formate		oz. .45			
" hippurate		oz. 2.00			
" **hydroxide** ("hydrate"), [hydrated (caustic) oxide], (Caustic Potassa), **chem. pure, Merck;**—an *absolutely pure* preparation,—free from Alumina, Silicic Acid, Sulphuric Acid, and Baryta		lb. 3.00			
" " pure (purif. *by Alcohol*), in sticks		lb. 1.10			
" " " (" "), in plates		lb. 1.05			
" " purified, in sticks,—*Potassa, U. S. Ph.*		lb. .65			
" " " in plates		lb. .60			
" " " in drops		lb. 1.25			
" " " dry, powder		lb. 1.50			
" " with Lime, [1:1], powder,—*Potassa cum Calce, U. S. Ph.*,—(Potassa-Lime)					
" " " " [2:1], powder, (Vienna Caustic Powder)					
" " " " [4:1], fused, (Filhos's Caustic; Fused Vienna Caustic)		lb. 2.00			
" hypo-phosphite,—U. S. Ph.		lb. 1.35			
" hypo-sulphite, see Potassium, thio-sulphate					
" imido-phtalate, (*Potassio-Phtal-imide*)					
" indigo-sulphate (sulph-indigotate, sulpho-cerulate)		oz. .75			
" iodate		oz. .55			
" iodide,—U. S. Ph. and Ph. G. II		lb. 3.75			
" iso-purpurate, chem. pure		oz. 5.00			
" lactate		oz. .50			
" lacto-phosphate (phospho-lactate)		oz. .55			
" manganate, (Mineral Chameleon—Chameleon Mineral)		lb. .40			
" methylo-sulphate		oz. .45			
" molybdate (molybdenate)		oz. .45			
" myronate		15 gr. 2.50			
" nitrate, chem. pure, cryst., (Refined Saltpetre), [Prismatic Nitre],—U. S. Ph. and Ph. G. II		lb. .50			
" " pure, powdered		lb. .50			

☞ When ordering, specify: "**MERCK'S**"!

	Containers incl.		
Potassium, nitrate, — (*continued !*); — in flat drops, (*tabulated*); [Tabulated Nitre; Prunella Salt]	lb. .65		
" do., with Zinc Chloride, fused; see under Zinc, chloride			
" nitrite, chem. pure, — in sticks	lb. 1.25		
" " commercial	lb. .75		
" nitro-prusside (nitro-prussiate; nitro-ferri-cyanide)	oz. 1.00		
" osmate, chem. pure	15 gr. 1.75		
" oxalate, neutral (normal), [so-called "sub-oxalate"], chem. pure	lb. .85		
" " " pure.—(Purity absolutely sufficient for photography.)	lb. .45		
N.B.—*Other oxalates:*—see Potassium: bin-oxalate; and, tetra-oxalate.			
" **oxide, hydrated (caustic),** [Caustic Potassa], chem. pure, Merck;—do., do., do., *U. S. Ph.*; and others, — see **Potassium, hydroxide, etc.**; etc.			
" per-chlorate	oz. .40		
" per-iodate	oz. 3.00		
" **per-manganate, pure, small cryst.,**—*U. S. Ph.*;—conforming to Ph. G. II.	lb. .50		
" " pure, large cryst.	lb. .55		
" " crude	lb. .40		
" phenate (phenylate, carbolate)	oz. .25		
" phosphate, pure, cryst.	lb. 1.25		
" " II, purified	lb. 1.15		
" phosphite	oz. .45		
" phospho-lactate, see Potassium, lacto-phosphate			
" plumbate	lb. 2.00		
" prussiates, so-called,—Red and Yellow,—etc., see Potassium: ferrid-cyanide, etc.; and, ferro-cyanide, *U. S. Ph.*, etc.			
" purpurate, Iso-, see Potassium, iso-purpurate			
" pyro-phosphate	oz. .35		
" quadro-oxalate, see Potassium, tetra-oxalate			
" rhodanide, see Potassium, sulpho-cyanate			
" ruthenate	15 gr. 4.00		
" salicylate	oz. .45		
" salicy*lite*	15 gr. 1.00		
" santoninate (*not* santonate!)	oz. 1.50		
" seleniate (selenate)	15 gr. .85		
" silicate, pure, dry	lb. 2.00		
" " " solution [10%]	lb. .50		
" " " " —sp. gr. 1.3.	lb. .75		
" " crude, solut. [30–33° Bé]	lb. .40		
" " " dry	lb. .50		
N. B.—*See, also:* Sodium, silicate. (Water-glass, Soluble Glass, or Liquid Glass.)			
" silico-fluoride	oz. .40		
" stannate	oz. .45		
" stearate	oz. 2.00		
" stibiate: Ph. Bor. VI; crude; and, pure; —see Potassium, antimonate: *pharmacopeial* (Ph. Bor. VI); do., do., *crude;* and, do., do., *pure*			
" stibiato - sulphide,—so-called,—see Potassa, antimonio-sulphurated, *crude*			
" succinate, neutral	oz. .65		
" sulphate, (Vitriolated Tartar), purified, cryst.	lb. .30		
" " purified, powder	lb. .30		
" " twice purified, cryst.	lb. .35		
" " " " powder	lb. .35		

☞ When ordering, specify: "**MERCK'S**"!

	Containers incl.			
Potassium, sulphate,—(*continued!*),—chem. pure, cryst.,—*U. S. Ph.* and Ph. G. II	lb. .60			
" do., do. do., powder................	lb. .60			
" sulphide, —*so-called,* —(Liver of Sulphur), crude, for baths;—and, purified,—*Potassa sulphurata, U. S. Ph.*; and, pure;—see Potassa, sulphurated, etc.; etc.; etc................				
" sulphite, normal.................	lb. 1.00			
" " " pure,—*U. S. Ph.*......	lb. 2.75			
" " acid, see Potassium, bi-sulphite..				
" sulpho-carbolate (sulpho-phenate, phenol-sulphonate)................	oz. .15			
" " -carbonate (thio-carbonate).—[An anti-phylloxerin].—(*See, also:* Potassium, xanthogenate.).....	lb. 1.50			
" " -cyanate (thio-cyanate; rhodanide), pure, cryst.............	oz. .24			
" " " commercial.............	oz. .20			
" " -indigotate (sulph-indigotate; sulpho-cerulate), see Potassium, indigo-sulphate................				
" " -vinate, see Potassium, ethylo-sulphate...................				
" tartrate, neutral, (Soluble Tartar), [Tartarus tartarisatus — Tartarized (Tartarated) Tartar], — cryst., pure,—Ph. G. H,—*Potassii tartras, U. S. Ph*...............	lb. 1.00			
" " do., powder, pure,—Ph. G. II...	lb. 1.05			
" " acid, see Potassium, bi-tartrate, *U. S. Ph.*; and other grades...				
" tellurite............................	15 gr. 2.50			
" tetra-oxalate (tetroxalate; quadro-oxalate), [sometimes—wrongly—called: "Essential Salt of Lemons"]......	lb. 3.00			
" thio-carbonate, see Potassium, sulpho-carbonate......................				
" thio-cyanate, see Pot., sulpho-cyanate..				
" thio-sulphate (formerly called "hyposulphite").......................	lb. 1.25			
" urate, pure.......................	oz. .80			
" valerianate.......................	oz. .75			
" wolframate (tungstate).............	lb. 2.00			
" xanthogenate (ethylo-thio-carbonate), I [An anti-phylloxerin.]—(*See, also:* Potassium, sulpho-carbonate.)	lb. 1.50			
" " II.....................	lb. 1.25			
Potassium and **Aluminium,** sulphate, see Alum, potassic...................				
" and **Ammonium,** fluoride;—readily soluble in Water.—(Emits fumes of Hydrofluoric Acid.)				
" " " phosphate.................	lb. 2.00			
" " " tartrate, (Ammoniated Soluble Tartar)....................	lb. 1.75			
" and **Antimony,** salts, see Antimony and Potassium...................				
" and **Barium,** chlorate, see Barium and Potassium, chlorate..............				
" and **Beryllium** (Glucinum), fluoride, see Beryllium and P., fluoride......				
" and **Bismuth,** salts, see Bism. and P..				
" and **Cadmium,** iodide, see Cadmium and Potassium, iodide.............				
" and **Chromium,** sulphate, see Alum, chromic.......................				
" and **Cobalt,** cyanide, see Potassium, cobalti-cyanide.................				
" and **Copper,** salts, see Copper and P...				
" and **Gold,** salts, see Gold and P......				

☞ When ordering, specify: "**MERCK'S**"!

	Containers incl.		
Potassium and **Iron,** cyanides, so-called, (Red and Yellow Prussiate of Potassa), etc.,—see Potassium : ferrid-cyanide, etc. ;—and, ferro-cyanide, *U.S.Ph.*; etc.			
" and **Iron,** ferro-cyanide, (Potassium ferri-ferro-cyanide; Soluble Prussian Blue), see Iron, cyanide, blue,—so-called,—*soluble*.....................			
" and **Iron,**—*other salts,*—see Iron, Mono-compounds; and, Iron, Sesqui-compounds, — (*the latter* embracing the *U. S. Ph.* Tartrate).................			
" and **Lithium,** tartrate, see Lithium and Potassium, tartrate.................			
" and **Mercury, salts;** see Merc. and P.			
" and **Nickel,** sulphate, see Nickel and Potassium, sulphate...............			
" and **Platinum,** double *and triple* salts, see Platinum double Chlorides; do. double Cyanides; do. *triple* Cyanides; and, do., divers double Salts........			
" and **Silver** Nitrates,—mixed in *U.-S.-Ph.* and other proportions,—(Mitigated Lunar Caustic), see Silver, nitrate, diluted, etc., etc............			
" and **Sodium,** boro-tartrate (tartaroborate), [Tartarus boraxatus—Borax-Tartar; *so-called* "Soluble Cream of Tartar"]....:....................	lb. 1.25		
" do. do., do.,—*in scales,*—(Scales of Borax-Tartar; "Soluble Scales of Tartar");—PERFECTLY SOLUBLE in Water, [*a property found wanting in other makes?*]	lb. 1.50		
" and **Sodium:** carbonate; and, sulphate;—see Sodium and Potassium, etc.; etc........................			
" and **Sodium,** tartrate, — (Tartarated [Tartarized] Soda; Soda-Tartar; Rochelle-salt, Seignette-salt), [Tartarus natronatus],—chem. pure, cryst.,—*U. S. Ph.* and Ph. G. II............	lb. .75		
" do. do., do.,—chem. pure, powder,—Ph. G. II.......................	lb. .80		
" and **Strontium,** chlorate, see Strontium and Potassium, chlorate.......			
" and **Titanium,** fluoride, see Titanium and Potassium, fluoride............			
" and **Zinc,** cyanide, cryst., see Zinc and Potassium, cyanide................			
" and **Zirconium,** fluoride, see Zirconium and Potassium, fluoride........			
Potassium, Lithium, and **Platinum,** cyanide } see under Platinum triple Cyanides ...			
" **Sodium,** and **Platinum,** cyanide..... }			
Potassium Alum, see Alum, potassic.....			
Powder, Blood, see Blood, bullock's, etc...			
" **James's,** (Febrile powder), see Antimonial Powder, *U. S. Ph.*...........			
" **Putty-,** so-called,—(*Polishing-powder,*) —see Tin, oxide, grey............			
" **Tin,** (Stanni pulvis), see Tin, metallic, pure, powder			
Powder of Algaroth, see Antimony, oxychloride			
Preparing-salt, so-called,—(Mordant),—see Sodium, stannate			
Primrose Yellow, see Aniline and Phenol Dyes: Yellow...................			

☞ When ordering, specify : "**MERCK'S**"!

	Containers incl.		
Propyl-amine, —10-% solution, aqueous " hydrochlorate " sulphate *These designations are frequently used erroneously, for the corresponding ones of:* "TRI-METHYL-AMINE," etc.,—*which see!*			
Propylene, bromide	oz. 2.00		
Protagon	15 gr. 3.00		
Protein	oz. 2.00		
Prunella Salt, see Potassium, nitrate, in flat drops			
Prussian Blue, ordinary, see Iron, cyanide, blue,—so-called,—*insoluble*			
do. do., soluble, see Iron, cyanide, blue,—so-called,—*soluble*			
Ptyalin, active	oz. .85		
Ptyalin-Pepsin	oz. 1.00		
Pulsatilla-camphor, see Anemonin			
Pulvis aërophorus cum *Magnesia citrica*, see Magnesium, citrate, effervescent. N.B.—*Compare, also:* Do., do., do., granulated, *U. S. Ph.*			
" **Sanguinis,** see Blood, bullock's, etc.			
" **Stanni,** see Tin, metallic, pure, powder			
Punicine (*not* Manna-sugar,—which is sometimes called "Punicin";—*but:* the Pomegranate Alkaloids!), see Pelletierine, etc.			
Purple of Alumina and Gold, see Gold, Alumina Purple of			
" **Cassius's,** see Gold, Tin-precipitate of.			
Purpurin, dry	oz. 1.50		
" paste,—free from Arsenic	oz. .40		
Putty-powder, so-called,—(Polishing-powder),—see Tin, oxide, grey			
Pyridine, chem. pure,—boiling-point 116–118° C [240.8–244.4 F]	oz. .30		
" nitrate, cryst.	oz. .75		
" sulphate, cryst.	oz. .75		
Pyro-catechin (Catechol; *ortho*-Di-oxybenzene)—[Pyro-catechuic (Oxy-phenic) Acid]	15 gr. .75		
Pyro-gallol, see Acid, pyro-gallic			
Pyro-gallol-phtalein, see Gallein			
Pyrolusite (Native Per-oxide of Manganese), see Manganese, oxide, black, *U. S. Ph.*			
Pyro-xylin, see Collodion Cotton			
Pyrrole (Pyrroline)	15 gr. .45		
" tetr-Iod-, see Iodole			

☞ When ordering, specify: "**MERCK'S**"!

	Containers incl.			

☞ When ordering, specify: "MERCK'S"!

Quassin, chem. pure, cryst.	Containers incl. 15 gr. .75			
" " " powder	⅛ oz.vls.oz. 6.00			
" purified, powder	⅛ oz.vls.oz. 4.00			
" " dry,—small lumps	⅛ oz.vls.oz. 3.50			
" sulphate, pure	15 gr. .50			
" —acc. to the French standard	⅛ oz.vls.oz. 2.00			
Quassin, Surinam, chem. pure, powder	15 gr. 2.50			
Quebracho Alkaloids:				
Aspido-spermine, cryst.,—acc. to Fraude	15 gr. 1.50			
" " " " sulphate	15 gr. 1.50			
Aspidos-amine, — acc. to Hesse	15 gr. 5.00			
" " " " hydrochlorate	15 gr. 5.00			
Quebrachine, cryst.,- acc. to Hesse				
" " " " " hydrochlorate	15 gr. 2.50			
Quebrach-amine,—acc. to Hesse	15 gr. 4.50			
" " " " sulphate	15 gr. 4.50			
Hypo-quebrachine, — acc. to Hesse	15 gr. 1.25			
" " " " hydrochlorate	15 gr. 1.25			
Aspido-spermine, pure,— *amorphous* ⎫ *commercial!*	15 gr. .75			
" citrate "	15 gr. 1.00			
" hydrochlorate "	15 gr. 1.00			
" sulphate " ⎭	15 gr. 1.00			
N. B. — These *commercial* (*amorphous*) Aspido-spermines are *not* homogeneous substances.				
Quercit (Acorn-sugar)	15 gr. .65			
Quercitrin. — Glucoside from Quercitron-bark—from Quercus tinctoria	15 gr. .35			
Quevenne's Iron, so-called, see Iron, metallic, reduced				
Quinetum (Quinio) [so-called "Mixed Alkaloids"—from Cinchona-bark],—pure	oz. 1.50			
" sulphate	oz. 2.25			
Quinidine (*Beta*-Quinidine[-Chinidine], *Beta-Quinine*, Beta-Chinine; Conchinine), —pure, cryst.	oz. .73			
" bi-sulphate	oz. .70			
" citrate	oz. .70			
" di-hydrobromate	oz. 1.75			
" hydrobromate	oz. 1.75			
" sulphate,—*U. S. Ph.*	oz. .33			
Quinidine, Alpha-, see Cinchonidine				
Quinine (Chinine; Quinia; *Alpha*-Quinine), pure,—*Quinina, U. S. Ph.*	oz. 1.20			
" acetate	oz. 1.20			
" æthylo-sulphate, see Quinine, ethylo-sulphate				
" ammonio-citrate, see Quinine and Ammonium, citrate				
" anisated, (Anethol-Quinine)	oz. 1.50			
" antimonate	oz. 1.35			
" arseniate (arsenate)	oz. 1.25			
" arsenite	oz. 1.50			
" benzoate	oz. 1.25			
" bi-muriate, carbamidated (*ureated*), see Quinine and Urea, hydrochlorate				
" bi-sulphate, *U. S. Ph.*, see Quinine, sulphate, acid				
" borate	oz. 1.40			
" " —*amorphous*, — see Quinoidine, borate				
" bromate	oz. 1.50			
" camphorate	oz. 1.50			
" carbolate, see Quinine, phenate				
" chinate, and chinovate ; see Quinine: quinate; and, quinovate				
" chlorate	oz. 2.00			

☞ When ordering, specify : "MERCK'S" !

MERCK'S INDEX.

	Containers incl.
Quinine — (continued!), — cinnamate (cinnamylate)	oz. 2.00
" citrate	oz. 1.05
" " with Ammonium Citrate, — *true double salt!* — see Quinine and Ammonium, citrate	
" " with Pyro-phosphate of Iron	oz. .50
" citrico-hydr. chlorate, see Quinine, hydrochloro-citrate	
" di-hydrobromate, ⎱ readily solu- ⎰	oz. 1.25
" di-hydrochlorate, ⎰ ble in Water. ⎱	oz. 1.50
" di-hydro-iodate (di-hydriodate)	oz. 2.00
" ethylo-sulphate (sulpho-vinate)	oz. 1.25
" ferri-arseniate (-arsenate)	oz. 2.00
" " -arsenite	oz. 1.50
" " -bromide	oz. 3.00
" " -citrate,—Ph.G.II,—[9-10% of anhydrous Quinine]; — free from Cinchonine	oz. .27
" " " —*Ferri et Quininæ citras*, U. S. Ph.,-[12% of anhydrous Quinine]	oz. .28
" " " —Ph. Neerl.;[13% anh. Q.]	oz. .28
" " " —Ph. Brit..[13.7% " "]	oz. .28
" " " —Ph. Ross..[13.4% " "]	oz. .28
" " " green......[10% " "]	oz. .35
" " " " ......[15% " "]	oz. .40
" " " " ......[20% " "]	oz. .45
" " " " ......[25% " "]	oz. .50
" " " with Strychnine, see under Strychnine	
" " -hydrochlorate (ferri-muriate)	oz. 2.50
" " -hydrocyanate	oz. 1.50
" " -hypo-phosphite	oz. 1.55
" " -iodide	oz. 1.55
" " -lactate	oz. 1.50
" " -muriate, see Quinine, ferri-hydrochlorate	
" " -sulphate	oz. 1.50
" " -tannate	oz. .75
" " -tartrate	oz. 1.25
" " -valerianate,—[33½% Quinine]	oz. 1.30
" formate	oz. 1.75
" hydrobromate,—*U. S. Ph.*	oz. 1.00
" hydrochlorate (muriate), cryst.,—*U.S.Ph.*	oz. .95
" " — *amorphous*, — see Quinoidine, hydrochlorate	
" " muriato-ureated (-*carbamidated*), see Quinine and Urea, hydrochlorate	
" hydrochloro-citrate, (citrico-hydrochlorate), cryst.—*A true double salt,*— slightly soluble in Water; more easily in Alcohol	oz. 2.50
" hydrofluorate	oz. 4.00
" hydro-iodate (hydriodate)	oz. 1.25
" hydro-silico-fluorate. — White microscopic crystals; little soluble in Alcohol; very readily soluble in Water	
" hypo-phosphite	oz. 1.55
" iodate	oz. 2.00
" kinate, and kinovate; see Quinine: quinate; and, quinovate	
" lactate	oz. 1.35
" lacto-phosphate (phospho-lactate)	oz. 2.00
" muriate, see Quinine, hydrochlorate	
" nitrate	oz. 2.00
" peptonized, (Peptone-Quinine)	oz. .75
" phenate (phenylate, carbolate), [Phenol-Quinine]	oz. 1.75

☞ **When ordering, specify: "MERCK'S"!**

	Containers incl.			
Quinine—(continued !),—phosphate	oz. 1.25			
" phospho-lactate, see Quinine, lacto-phosphate				
" phtalate. — Light, translucent scales; perfectly soluble in 2 parts of 95-% Alcohol;—this solution, with proper care, is dilutable by Water.—Melting-point 70° C [158 F]	oz. 2.00			
" picrate	oz. 2.00			
" quinate (chinate, kinate)	oz. 3.00			
" quinovate (chinovate, kinovate)	oz. 3.00			
" saccharinate (not saccha-rate !) } *True salts of Quinine and Saccharin — which latter see !*				
" " bi-				
" salicylate	oz. 1.10			
" santoninate (not santonate!)	oz. 6.00			
" stearate (stearinate)	oz. 1.50			
" stibiate, see Quinine, antimonate				
" succinate	oz. 1.75			
" **sulphate, pure, neutral, — Zimmer's :** — in $1/16$-, $1/8$-, $1/4$-, $1/2$-, and 1-oz. vials ; and in 1-, 5-, 10-, 25-, 50-, and 100-oz. tins	} [*Regarding prices, see remark on page 158 !*]			
" " chem. pure,— U. S. Ph.,—made from the Bi-sulphate	oz. .65			
" sulphate, acid, (bi-sulphate,—U.S.Ph.),—[about 60% Quinine]	oz. .55			
" sulpho-carbolate (phenol-sulphonate, sulpho-phenate), cryst.	oz. 2.00			
" sulpho-vinate, see Quinine, ethylo-sulphate				
" sulphurico-tartrate (tartarico-sulphate)	oz. 2.00			
" tannate, commercial	oz. .55			
" " Ph. G. I,—[20% pure Quinine]	oz. .75			
" tannate, neutral, true,—insipid	oz. 1.00			
" tartarico-sulphate, see Quinine, sulphurico-tartrate				
" tartrate, cryst.	oz. 1.25			
" thymate	oz. 5.00			
" urate	oz. 2.50			
" valerianate,—U. S. Ph.,—large cryst.; —free from Cinchonidine	oz. 1.30			
Quinine and **Ammonium,** citrate, (Ammonio-citrate of Quinine),—true double salt.—Slightly soluble in Water; more easily so in Alcohol				
" and **Urea,** hydrochlorate, (Ureated [carbamidated] Di-hydrochlorate of Quinine; Muriato-carbamidated Hydrochlorate of Quinine)	oz. 2.00			
Quinine-**Iron salts,** see "Quinine, ferri-—," etc.,—(above!)				
Quinine, **Anethol-,** see Quinine, anisated				
" **Peptone-,** see Quinine, peptonized				
" **Phenol-,** see Quinine, phenate				
Quinine, amorphous, true, see Quinoidine				
" do., so-called, see Quinium Labarraque				
Quinine, **Alpha-,** see Quinine				
" **Beta-,** see Quinidine				
Quinine-flower (Quinine Plant), so-called; —Glucoside from,—see Sabbatin				
Quinio,—and do., sulphate,—see Quinetum, etc.				
Quinium Labarraque, (Chinium), [Alcoholo-calcic Extract of Cinchona-bark;—so-called "Amorphous Quinine"]	oz. .75			
Quinoidine (Chinoidine — *Chinoidina!*), —[*True* Amorphous Quinine],—pure	oz. .15			
" chem. pure,—Ph. G. II;—the so-called "*Chinoidinum*" of the U. S. Ph.	oz. .16			

☞ **When ordering, specify : " MERCK'S " !**

	Containers incl.			
Quinoidine—(*as above!*),—borate, (Borate of Amorphous Quinine)...............	oz. .35			
" citrate, in scales...................	oz. .30			
" hydrochlorate, (Hydrochlorate of Amorphous Quinine)....................	oz. .50			
" sulphate, dry......................	oz. .25			
" tannate	oz. .30			
Quino-iodine (Chino-iodine — *Chino-iodinum!*).—[*Do not confound* with: *Quinoidine*, —(above!).]............................				
Quinoline, (Quinoleine; **Chinoline**, Chinoleine)— [**Leucoline**, Leucol], —**synthetical** (= **medicinal!**),—**chem. pure**;—boiling-point 230–234° C [446–453.2 F].....	oz. 1.50			
" —synthetical (= medicinal!),—pure ...	oz. .50			
" —do.,—citrate.....................	oz. 1.00			
" " ferri-citrate,—[10%]	oz. .75			
" " " —[20%].................	oz. .85			
" " hydrochlorate.................	oz. 1.50			
" " salicylate.....................	oz. 1.00			
" " sulphate	oz. 1.50			
" " tannate	oz. 1.00			
" " tartrate, pure, perf. white,—non-deliquescent.................	oz. .50			
Quinoline Blue, (Chinoline-iodo-cyanine), see **Cyanine**...............................				
Quinoline-Hydro-quinone (Chinoline-Hydro-chinone), **cryst.**.....................	oz. 3.00			
" **-Resorcin** (Chinoline-Resorcin).........	oz. 2.50			
Quinone (Chinone) [Benzene-(Benzol-, Benzo-)Quinone]—(Chinoyl)..................	oz. 5.00			
Quinone Hydride, see **Hydro-quinone**..........				

N. B.—Other *Cinchona derivatives* than above named under "**Q**",—see Cinchonidine, Cinchonine;—see, also: Acid, quinic; do., quino-picric; do., quinovic.—(Also: *some salts* of these Acids,—under the names of the *Metals* or *Radicles* of their respective bases.)

When ordering, specify: "**MERCK'S**"!

	Containers incl.			
Reagent Papers, see Paper, etc...........				
Realgar, see Arsenic, Red sulphide.........				
Regulus of **Antimony,** see Antimony, metallic..................................				
Rennet - powder, I, — (coagulates 100,000 parts of milk)......................				
" II, — (coagulates 20,000 parts of milk)..				
Rennet Wine, (*Liquid Rennet*), [Liquor seriparus; so-called "Essence" of Whey] ...				
Resineon...............................	oz. .35			
Resins (Resinæ):				
Brayera, see Resin, Kousso...............				
Copaiva, — (Balsamum copaivæ siccum), [Crude Copaivic Acid]..................	lb. 1.25			
Indian Hemp, (Cannabis indica)..........	oz. 1.00			
Jalap, — brown: from the *true* root (Tuber of Ipomœa purga [Exogonium purga]); — consists principally of Convolvulin — (*which see also!*).............	oz. .35			
" —do.: as above, — Ph. G. II............	oz. .50			
" —white: from the *true* root; — (*the pure Glucoside!*): — see Convolvulin.....				
" —brown: from the *light* root (Orizaba root; Male [Fusiform] Jalap, — from Convolvulus orizabensis); — consists principally of Jalapin — [*which see also!*]...........................				
" —white: from the *light* root; — (*the pure Glucoside!*): — see Jalapin				
Kamala (Glandulæ Rottleræ tinctoriæ)....	oz. 1.00			
Kava-Kava (Ava), [Radix macropiperis], **Alpha-**	15 gr. .50			
" **Beta-**	15 gr. .25			
" *both* the above mixed, in proportion as contained in the root................	15 gr. .40			
Kousso (Koosso, Cusso) [Brayera]: flowers	oz. 3.00			
Mezereon (Daphne mezereum — Spurge Olive): bark........................	oz. 1.50			
Quebracho blanco, (White Quebracho): bark	oz. 3.50			
Scammony: root, — Ph. G. I; — consists essentially of Scammonin — (*which see also!*) — and which is identical with Jalapin).........................	oz. .75			
" do., — white; (*the pure Glucoside!*), — in sticks or powder, — see Scammonin.				
Spurge Olive, see Resin, Mezereon........				
Sumbuli-root (Musk-root)................	oz. 3.50			
Turpeth-root, — (= Turpethin)............	oz. 1.50			
Veratrum, Green, (Indian Poke), [*American Green Hellebore*].....................	oz. 2.00			
Resorcin, (Resorcinol), [meta-Di-oxy-benzene], chem. pure, cryst., perfectly white......	oz. .30			
" chem. pure, resublimed, perfectly white...	oz. .70			
" chem. pure, impalpable powder, — for dry-spray atomization. — (Escharotic inhalant.)	oz. .85			
Resorcin, di-, see Di-resorcin.............				
Resorcin, Pheno-, see Pheno-Resorcin ...				
Resorcin-phtalein, see Fluorescin				
" -phtalin, see Fluorescin				
Rhabarbarin, } see under Rhubarb constituents Rhein........ }				
Rhodium, metallic.....................	15 gr. 5.00			
Rhubarb (*Rheum*) **constituents:**				
Erythro-retin (Rhabarbarin).............	15 gr. .50			
Rhein, cryst., — (*True* Chrysophanic Acid; Rheic Acid), [*Rhubarb Yellow*]..........	15 gr. 1.50			
N.B. — *So-called* "**Medicinal Chrysophanic Acid,**" see Chrys-arobin.				
Ricinine				

☞ When ordering, specify: "**MERCK'S**"!

	Containers incl.
Rochelle (Seignette) Salt, see Potassium and Sodium, tartrate, *U. S. Ph.;* etc......	
Ros-aniline, hydrate.....................	oz. .75
" acetate.............................	oz. .75
" hydrochlorate	oz. .75
" iodide............................:	
Ros-aniline-sulphonate of Sodium, see Sodium, ros-aniline-sulphonate...........	
Ros-aurin, see Acid, rosolic..............	
Rotoin,—from Japanese Belladonna, (Scopolia japonica).........................	15 gr. 2.50
Rubidium, metallic, pure	15 gr. 20.00
" bi-tartrate, cryst.....................	15 gr. .75
" chloride...................:	15 gr. .50
" iodide	15 gr. 1.00
" sulphate	15 gr. .65
Rubidium and Caesium, chloride, see Caesium and Rubidium, chloride.........	
Rubidium Alum, see Alum, rubidic.......	
Rubigo, see Iron, oxide, brown, *pure*......	
Ruthenium, metallic....................	15 gr. 5.50

☞ **Reagents, Merck's guaranteed,**—*for analyses;*—see page 155!

☞ When ordering, specify: "**MERCK'S**"!

	Containers incl.
Sabadilline, pure.......................	15 gr. .75
" sulphate............................	15 gr. .75
Sabbatin.—Glucoside from Sabbatia Elliottii—the so-called "Quinine Plant," or "Quinine-flower"............................	15 gr..1.50
Saccharated Iron, so-called, see Iron, oxide, red, saccharated....................	
" **Iron-salts,** *divers,* see *references under:* Sugar, ferruginated; *or under:* Iron, saccharate; *or under:* Iron-Sugar....	
" **metallic Salts,** *divers,* see under the names of the respective metals......	
Saccharin Fahlberg, *(not a Carbo-hydrate,* but : ortho-Sulph-amine-benzoic Anhydride!).—[*Non-fermentable* sweetening agent, of 280-fold the intensity of Cane-sugar.]— (Anti-zymotic;—of high importance in diabetes, gastric disorders, etc.)............	oz. 1.25
☞ N. B.—*See, also:* the *Saccharinates* and *Bi-saccharinates* of Morphine, Quinine, and Strychnine, (*under these* Alkaloids). —Those *true Salts—not to be confounded* with Sugar-compounds [so-called "Saccharates"]!—are useful when the *taste* of bitter Alkaloids is to be disguised.	
Saccharum Carnis, (Meat-sugar), see Inosit	
" **Lactis,** see Milk-sugar...............	
" **Mannæ,** see Mannit..................	
" **Plumbi** (*Saturni*), see Lead, acetate, normal, *U. S. Ph.;* and other grades...	
" **Seminis Quercus,** (Acorn-sugar), see Quercit............................	
" **uveum** (*amylaceum*), see Grape-sugar..	
N.B.—*Other Sacchara*, see under Sugar.	
Safflower Carmine......................	oz. 2.50
Saffron (Crocus) **of Antimony,** [Crocus metallorum], see Potassa, antimoniosulphurated, *washed*................	
" **of Iron,** (Crocus martis),—*aperient*,—see Iron, oxide, brown, (so-called sub-carbonate)............	
" " " —*astringent*,—see Iron, oxide, red, *anhydrous*............	
Safranine, see under Aniline and Phenol Dyes:—Red; and, Yellow................	
Safrol,—sp. gr. 1.108......................	lb. 1.00
Sal Acetosellæ, see Potassium, bin-oxalate.	
" **amarum,** see Magnesium, sulphate, *U. S. Ph.;* and other grades and forms...	
" **ammoniacum,** see Ammonium, chloride, *U. S. Ph.;* and various other kinds	
Sal Soda, see Soda, carbonate, neutral, *U. S. Ph.s;* and other grades and forms........	
Sal, etc.,—*other than above,*—see Salt, etc...	
Salicin,—*U. S. Ph.*......................	lb. 2.75
Salicyl-Resorcin-ketone (-acetone), [Trioxy-benzo-phenone].....................	15 gr. .75
Salicylal (Salicylol) [Salicyl Hydride; *Salicylic Aldehyd*], see Acid, salicylous........	
Saligenin (ortho-Oxy-benzylic Alcohol ; Salicylous Alcohol).........................	15 gr. .50
Sali-naphthol, see **Betol**...................	
Salithol, see Phenetol......................	
Salol (Phenylic Ether of Salicylic Acid; Salicylate of Phenol).....................	oz. .40
Salt, Dyers', (Pink Salt), see Tin and Ammonium, chloride.....................	
" **Epsom,** see Magnesium, sulphate, *U. S. Ph.;* and other grades and forms....	

☞ **When ordering, specify : "MERCK'S" !**

	Containers incl.		
Salt, Figuier's, of Gold, see Gold and Sodium, chloride, cryst.			
" Glauber's, see Sodium, sulphate, *U. S. Ph.*; and other grades and forms....			
" Gregory's, (Hydrochlorate of Morphine and Codeine).....................	⅛ oz.vls.oz. 5.00		
" Karlsbad thermal, artificial, large cryst.	lb. .12		
" " " " small cryst. ...	lb. .12		
" " " " dry,—Ph.G.II.	lb. .25		
" " " true...............	lb. 1.75		
" Kreuznach, (the German "Kreuznacher Mutterlaugensalz")................	lb. .12		
" Magnus's "green," see Platinum double Chlorides: Platinum-tetr-amine and Platinum, bi-chloride			
" microcosmic, see Sodium and Ammonium, phosphate·..................			
" Monsel's, see Iron, sub-sulphate......			
" mordant, see Sodium, stannate			
" pink (Dyers'), see Tin and Ammonium, chloride.........................			
" preparing-, so-called,—(Mordant Salt), see Sodium, stannate...............			
" Prunella, see Potassium, nitrate, in flat drops..........................			
" Rochelle (*Seignette*), see Potassium and Sodium, tartrate, *U. S. Ph.*; etc......			
Salt of Amber, volatile, see Acid, succinic			
" of Gold, Figuier's, see Gold and Sodium, chloride, cryst.			
" of Lemons,—Essential,—(so-called), —see Potassium, bin-oxalate; etc.; —and also: tetra-oxalate............			
" of Sorrel, see Potassium, bin-oxalate .			
" of Tartar, see Potassium, carbonate, *pure, U. S. Ph.*; and other grades....			
" of Tartar,—Essential,—see Acid, tartaric, *U. S. Ph.*; and other kinds			
" of Tin,—so-called, — anhydrous, see Tin, chloride			
Saltpetre, refined, see Potassium, nitrate ..			
" Soda-, see Sodium, nitrate...........			
Sanguinarine, pure.....................	15 gr. 1.00		
" nitrate	15 gr. 1.00		
" sulphate	15 gr. 1.00		
Sanguis *Tauri* (*Bovis*) siccus pulveratus, see Blood, bullock's, etc.....................			
Santalin (Santalic Acid)....................	oz. .85		
Santonin,—*U. S. Ph.*,—(Anhydride of Santoninic [*not* Santonic!] Acid);— [$C_{15}H_{18}O_3$],—cryst.	oz. .45		
" powder	oz. .45		
N.B.—*See, also:* Acid, santoninic.			
Sapo, see Soap......................			
Saponin, pure,—*from Saponaria officinalis.*— (Chemically identical with Senegin [Polygalin],—from Senega.).........	⅛ oz.vls.oz. 2.00		
" crude...............................	oz. .40		
Sapo-toxin, — acc. to Kobert. — Fractional derivative of Saponin *from the bark of Quillaia saponaria;* — a white, amorphous, non-crystallizable powder; easily soluble in Water.—(An intensive heart-poison.) ...	15 gr. .75		
Sarcine (Hypo-xanthine)...................	15 gr. 5.00		
" hydrochlorate.....................	15 gr. 5.00		
Sarcosine (Methyl-glycocoll [-glycocine])..	15 gr. 6.00		
Sarsaparin (Parillin), see Smilacin........			
Scales of Tartar (-of *Borax-Tartar*), soluble (*perfectly soluble* in Water);—see Potassium and Sodium, boro-tartrate,—*in scales*			

☞ When ordering, specify: "MERCK'S"!

MERCK'S INDEX.

	Containers incl.			
Scammonin (White Resin of Scammony), —*the pure Glucoside;*—[identical with JALAPIN; *but* from the root of Convolvulus scammonia];—in sticks....	oz. .80			
" —in powder........................	oz. .85			
N.B.—*See, also:*—Resins; Scammony, root, —Ph. G. I.				
Scilla preparations,—(Scilli-picrin, Soilli-toxin, Scillitin),—see Squill preparations..........				
Scoparin (Scoparic Acid).................	15 gr. .65			
Scopoleine.=Alkaloid from Japanese Belladonna, (from Scopolia Japonica)..........	15 gr. 3.50			
Seignette (Rochelle) Salt, see Potassium and Sodium, tartrate, *U. S. Ph.;* etc.......				
Selenium, in sticks......................	oz. 3.00			
" — in the form of a Berzelius medallion	each 4.00			
" hydroxide, Selenic, (Hydrated Tri-oxide), see Acid, selenic................				
" oxide, Selenious, (Di-oxide), sublimed, see Acid, selenious, anhydrous........				
Senegin (Polygalic Acid, Polygalin),—*from Senega.*—[Chemically identical with Saponin,—from Saponaria officinalis.]..........	15 gr. .75			
Senna-leaves, de-resinated,—powdered...				
Sero-sublimate (Serum, with Corrosive Sublimate),—[1%],—liquid;—according to Lister............................	lb. 1.50			
" —in scales;—according to Lister......	oz. .75			
Silica (*Silicea; Silex*), pure, see Acid, silicic.				
Silicon (*Silicium*), so-called "metallic". cryst.	15 gr. 2.25			
" do. "do.," amorphous...............	15 gr. 1.75			
" bromide............................	15 gr. .40			
" chloride...........................	15 gr. .35			
" di-oxide, (Silicic Oxide), see Acid, silicic				
Silver (Argentum), double salts of, see "Silver and —" (below!).................				
" metallic, precipitated, powder........	oz. 4.00			
" acetate, chem. pure..................	oz. 2.50			
" albuminate.........................	oz. 2.50			
" ammonio-fluoride. } see Silver and Ammonio-nitrate { monium, etc.; etc.				
" arsenite............................	oz. 2.50			
" borate.............................	oz. 2.50			
" bromide...........................	oz. 2.00			
" carbonate..........................	oz. 3.00			
" chloride...........................	oz. 1.50			
" chromate..........................	oz. 2.50			
" cyanide,—*U. S. Ph.*.................	oz. 2.50			
" fluoride, ammonio-, see Silver and Ammonium, fluoride.				
" iodide,—*U. S. Ph.*...................	oz. 3.00			
" lactate.............................	oz. 4.00			
" mono-chlor-acetate, cryst............	oz. 6.00			
" nitrate, cryst.,—*U.S.Ph.*,—(Lunar Nitre).......................... ⎫ ⎪ [Lapis Infernalis Internal Stone]	oz. 1.25			
" " molded (fused),—*U.S. Ph.*,—prf.colorless. ⎬ [Lunar Caustic.]	oz. 1.22			
" " do., grey....... ⎪	oz. 1.25			
" " " " -pencils, in woodencase ⎭	doz. 1.22			
" nitrate, diluted, (*with Potassium Nitrate* —1:1),—*U. S. Ph.*,—[Mitigated (toughened) Caustic],—sticks.......	oz. 1.00			
" " do. (do. do. do.),—in the following proportions [of Silver Nitrate to Potassium Nitrate]:—				
1:2; sticks,—Ph. G. I & II..	oz. .75			
1:3; "	oz. .50			
1:4; "	oz. .50			
1:5; "	oz. .55			

☞ When ordering, specify: "MERCK'S"!

	Containers incl.			
Silver, nitrate, diluted,—(*as above!*); in the following proportions [of Silver Nitrate to Potassium Nitrate],—*continued:*—				
2:1; sticks	oz. 1.10			
2%; sharpened pencils,—sizes *as follows:*—				
No. of pieces. Weight abt. gm. Long cm. Thick mm.				
4 = 30; ea. 7 5	oz. 1.50			
6 = 30; " 5.5...5	oz. 1.55			
8 = 30; " 6 ....3.5	oz. 1.60			
" nitrate, with Silver Chloride—[10%]	oz. 2.50			
" " " Lead Nitrate,—[5:1]	oz. 2.50			
" nitrate, ammonio-, see Silver and Ammonium, nitrate				
" nitrite	oz. 2.50			
" oleate	oz. 2.50			
" oxalate	oz. 2.75			
" oxide,—*U. S. Ph.*,—(Argentic Oxide, Mon-oxide)	oz. 2.75			
" per-manganate, pure	oz, 2 50			
" phosphate	oz. 2.25			
" silvate (silvinate)	oz. 4.00			
" sulphate, cryst	oz. 1.75			
" sulphide (sulphuret)	oz. 3.50			
" tartrate	oz. 2.25			
" tri-chlor-carbolate (tri-chlor-phenate)	oz. 2.25			
Silver and Ammonium, fluoride.—(Used in Chromo-photography.)				
" and **do.**, nitrate	oz. 2.50			
" and **Potassium** Nitrates,—mixed in *U.-S.-Ph.* and other proportions, (Mitigated Lunar Caustic), see Silver, nitrate, diluted, etc.; etc.				
" and **Sodium,** thio-sulphate (formerly called "hypo-sulphite")	oz. 3.00			
Simulo,—see under Tinctures				
Skatole	15 gr. 6.00			
Smilacin (Parillin, Pariglin, Sarsaparin), cryst.	15 gr. 1.75			
Snail-juice, saccharated, see Helicina				
Soap (Sapo), butyric (of Butter),—for preparing Opodeldoc.	lb. .40			
" of Castor-oil and Magnesia, (Sapo ricini magnesicus), [Ricinated Magnesia], see Magnesium, ricinate				
" medicinal, powder ⎰ Sapo, ⎱	lb. .60			
" " in bars .... ⎱ *U.S.Ph.* ⎰	lb. .15			
" " —Ph. G. II,—powder	lb. .75			
" " " —in bars	lb. .20			
" green (soft) [potassic],—*Sapo viridis, U. S. Ph.*,—Sapo kalinus, Ph. G. II	lb. .25			
" Castile (hard),—Sapo venetus [oleaceus, hispanicus]	lb. .15			
Soda (Natrum, Natron), caustic, see Sodium, hydroxide, etc.; etc.				
" *U. S. Ph.*,—see Sodium, hydroxide, pure (*purif. by Alcohol*); sticks				
Soda, sulphurated,—(Sodic Liver of Sulphur), [improperly called "Sodium Ter-sulphide"],—fused ..	lb. .85			
" " —fused, pure	lb. 1.25			
N.B.—*Compare, also:* Sodium, sulphide (sulphuret), cryst., *true.*				
Soda, tartarated (*tartarized*), [Soda-Tartar], see Potassium and Sodium, tartrate, *U. S. Ph.*; etc.				
Soda Alum, see Alum, sodic				
Soda-Lime, see Sodium, hydroxide, with Lime				
Soda Saltpetre, see Sodium, nitrate				

☞ **When ordering, specify: "MERCK'S"!**

	Containers incl.			
Soda-Tartar (Tartarated [Tartarized] Soda), see Potassium and Sodium, tartrate, *U. S. Ph.*; etc.				
Sodio-Ethyl (Natrio-Ethyl), see Sodium, ethylate, etc.; etc.; etc..................				
Sodium (Natrium), double and triple salts of, see "Sodium and —" (below!)...				
" metallic.............................	lb. 3.50			
" acetate, cryst., (Terra foliata tartari *crystallisata*).............	.lb. .45			
" " " chem. pure,—*U. S. Ph*...	lb. .75			
" " " pure, fused...............	lb. .85			
" aceto-wolframate (aceto-tungstate).....	lb. 1.25			
" æthylate, see Sodium, ethylate........				
" æthylo-sulphate, see Sodium, ethylo-sulphate............................				
" antimonate, Meta-, see Sodium, meta-antimonate...................				
" " Pyro-, see Sodium, pyro-antimonate............................				
" arseniate (arsenate), di-sodic, dry......	lb. .60			
" " do., cryst., — *Sodii arsenias, U. S. Ph.*....................	lb. .35			
" " " pure.....................	oz. .14			
" arsenite............................	lb. .50			
" " pure	oz. .14			
" benzoate,—*U. S. Ph.*,—from artificial Benzoic Acid................	oz. .24			
" " from true Benzoic Acid from the resin..................	oz. .30			
" benzoico-sulphite, so-called, see Sodium, sulphite, benzoated.................				
" . bi-borate (pyro-borate, di-meta-borate), [Borax; Officinal Borate of Sodium], — fused; — (*Borax-glass*, Vitrified Borax)...............	lb. 1.50			
" " calcined, (Burnt Borax).........	lb. .75			
" " pure, cryst., prismatic (with 10 molecules of Water),—*U. S. Ph.*; —(Refined Borax)............	lb. .75			
" " cryst., prismatic, (Crude Borax).	lb. .40			
" " powder,—*from prismatic crystals*, —(*not* Amorphous Borax!).....	lb. .50			
" " —glycerolate of, ("Glycerite" of Borax—*Glyceritum Sodii boratis, U. S. Ph.* 1870;— Glycerinum Boracis, Ph. Br.),—[1 part Borax; 4 Glycerin; 2 Water]...				
' "' —do. do., *-syrupy consistency,*-(improperly called : "Boro-Glycerin"),—[about *equal parts* Borax and Glycerin] ;—(*not to be confounded with the true—Dry* —Boro-Glycerin = Glycerolate of Boric Acid!).............. N.B.—*See, also:* Boro-Glycerin.	lb. 1.50			
" bi-carbonate (acid carbonate; hydro-carbonate), chem. pure, cryst., in crusts	lb. .40			
" " chem. pure, cryst., in lumps.....	lb. .40			
" " " " powd.,—*Sodii bicarbonas, U. S. Ph...*	lb. .35			
" " pure, powder, — *Sodii bicarbonas venalis, U. S. Ph*..............	lb. .30			
" " English,—powder..............	lb. .30			
" " " " —in lumps...............	lb. .25			
" bi-chromate..........................	lb. .35			
" bin-oxalate..........................	lb. .75			
" bi-phosphate........................	lb. 1.25			

☞ **When ordering, specify: "MERCK'S"!**

		Containers incl.			
Sodium, bi-sulphate (acid sulphate), [Sodium and Hydrogen, sulphate], pure, cryst.		lb.	.60		
" do., pure, fused......................		lb.	.65		
" " do., do., — in drops. — Clearly soluble in Water.—(Decomposes carbonates, and is therefore employed for the production of Pure Carbonic Anhydride.)..........					
" " crude.............................		lb.	.30		
" bi-sulphite, dry, commercial, II ⎞ ⎰An Anti-⎱		lb.	.50		
" " solution, comm'l, -[30°Bé] ⎬ ⎱chlor! ⎰		lb.	.40		
" " dry, pure,— U. S. Ph.... ⎠		lb.	.60	•	
N.B.—See, also (for "Antichlor"):— Sodium, sulphite; and: do., thio-sulphate.					
" bi-tartrate, cryst......................		lb.	1.25		
" bi-vanadate, cryst.,—readily soluble...					
" borate, (Borax), see Sodium, bi-borate, U. S. Ph.; and other forms and grades					
" boro-benzoate		oz.	.50		
" " -citrate		lb.	2.00		
" " -salicylate.....................		oz.	.40		
" bromate		oz.	1.00		
" bromide,— U. S. Ph. and Ph. G. II....		lb.	.90		
" butyrate		lb.	2.00		
" camphorate		oz.	1.25		
" carbolate, see Sodium, phenate........					
" carbonate, neutral,— (Sal Soda),— twice purified, cryst.................		lb.	.25		
" " do.,—twice purified, dry		lb.	.35		
" " " ch. pure, cryst.,—U. S. Ph. and Ph. G. II..		lb.	.40		
" " " " " dried,— U. S. Ph.		lb.	.50		
" " " " " dry (anhydrous)..		lb.	.75		
" " " " " fused		lb.	1.25		
" carbonate, acid, see Sodium, bi-carbonate, U. S. Ph.s; and various others..					
" caustic oxide, — U. S. Ph.; and other grades,—see Sodium, hydroxide, etc.					
" chlorate, cryst.—U. S. Ph.....:......		lb.	.60		
" chlorhydro-phosphate, see Sodium, phosphate, hydrochlorated.........		oz.	.40		
" chloride, chem. pure, cryst.,— U. S. Ph.		lb.	.40		
" " " " exsiccated (decrepitated)..........		lb.	.50		
" " " " fused.............		lb.	.65		
" choleate (choleinate), pure,—Ph.G.I,— [Dried purified Ox Gall].......		oz.	.35		
" " —from Choleic (Tauro-cholic) Acid,—see Sodium, tauro-cholate.					
" chromate, neutral....		lb.	.40		
" " " pure		lb.	2.00		
" cinnamate, (cinnamylate), chem. pure..		oz.	2.00		•
" citrate, acid		lb.	2.00		
" " neutral		lb.	1.75		
" citrico-benzoate,—very freely soluble..		oz.	.65		
" copaivate		oz.	1.00		
" cresotate		oz.	.70		
" cyanide		oz.	1.25		
" **di-iod-para-phenol-sulphonate**, see **Sozoiodole**					
" di-meta-borate, see Sodium, bi-borate..					
" di-nitro-cresylate		oz.	1.50		
" ethylate, (Sodio-[Natrio-]Ethyl), dry...		oz.	1.00		
" " cryst., (Caustic Alcohol),—acc. to Richardson		oz.	.40		
" " liquid, (Liquor Sodii ethylatis), —Ph. Brit. ....•.......		lb.	2.00		
" ethylo-sulphate (sulpho-vinate), chem. pure............................		lb.	1.50		

☞ When ordering, specify : "**MERCK'S**"!

		Containers incl.		
Sodium, ethylo-thio-carbonate, see Sodium, xanthogenate...............				
" ferro-cyanide, (Sodio-Ferrous cyanide, so-called), pure............	oz.	.50		
" " commercial...............	lb.	.75		
" fluoride, pure..................	oz.	.45		
" " commercial...............	oz.	.25		
" formate, pure, dry.............	oz.	.50		
" glycerino-borate, (Glycerolate of Borax—*Glyceritum Sodii boratis*, U. S. Ph.1870), see Sodium, bi-borate, glycerolate of. .N.B.—*See, also:* Do., do., do. do.,—*syrupy consistency.*				
" glyco-cholate, cryst..............	15 gr.	1.50		
" **hippurate**...................	oz.	2.00		
" hydro-carbonate, see Sodium, bi-carbonate................				
" hydrochloro-phosphate, see Sodium, phosphate, hydrochlorated........				
" hydrogenio-sulphate, see Sodium, bi-sulphate................				
" hydrophosphate, (Di-sodium Hydroph.), see Sodium, phosphate, bi-basic....				
" hydroxide ("hydrate") [hydrated (caustic) oxide], (Caustic Soda), chem. pure,—from Sodium........	lb.	5.00		
" " pure (*purif. by Alcohol*); plates...	lb.	1.05		
" " " (" " "); sticks,-*Soda, U. S. Ph.*.	lb.	1.09		
" " purified, dry...............	lb.	.60		
" " " —in plates.............	lb.	.50		
" " " —in sticks.............	lb.	.55		
" " " —in drops.............	lb.	1.50		
" " crude,—[abt. 75%]............				
" " with Lime,—(Soda-Lime).......	lb.	.60		
" hypo-phosphite,—*U. S. Ph.*.........	lb.	1.30		
" hypo-sulphate, chem. pure..........	oz.	1.00		
" hypo-sulphite (sub-sulphite),—[*an Anti-chlor!*], see Sodium, thio-sulphate................				
" " chem. pure,—*U. S. Ph.*,—see do. do., chem. pure..........				
" **ichthyol-sulphonate** (sulpho-ichthyolate), see under **Ichthyol preparations**......				
" indigo-sulphate (sulph-indigotate, sulpho-cerulate), chem. pure.........	oz.	1.50		
" iodate.....................	oz.	1.00		
" iodide, dry,—*U. S. Ph.* and Ph. G. II..	oz.	.35		
" kousseinate.................	15 gr.	.50		
" lactate,—syrupy consistency.—(*N. B.—This consistency is the only form in which pure Sodium Lactate is obtainable.*).................	oz.	.35		
" lacto-phosphate (phospho-lactate). ...	oz.	.50		
" meta-antimonate (-stibiate), pure, cryst.	oz.	.40		
" meta-phosphate...............	oz.	.45		
" methylo-sulphate, cryst...........	oz.	.50		
" *methyl-tri-hydro-oxy-quinoline-carbonate*, see Thermifugin............				
" molybdate (molybdenate)...........	oz.	.50		
" muriato-phosphate, see Sodium, phosphate, hydrochlorated...........				
" nitrate, crude.......... ⎱ Soda Salt- ⎰				
" " purified....... ⎱ petre; Cu- ⎰	lb.	.35		
" " ch.pure,—*U.S.Ph.* ⎰ bic Nitre. ⎱				
" and Ph. G. II... ⎰	lb.	.50		
" nitrite, chem. pure,—in sticks........	oz.	.22		
" " commercial, - cryst...........	lb.	.40		
" nitro-prusside (nitro-prussiate; nitroferri-cyanide)................	oz.	1.00		

☞ When ordering, specify: "**MERCK'S**"!

	Containers incl.			
Sodium, oleate............................	lb. 1.50			
" ortho-phosphate, di-sodic, see Sodium, phosphate, bi-basic................				
" osmate, chem. pure..................	15 gr. 2.50			
" oxalate...............................	lb. .75			
" " chem. pure.....................	lb. 1.00			
" oxide, hydrated (caustic), [Caustic Soda], — *U. S. Ph.*; and other grades and forms,—see Sodium, hydroxide, etc.; etc..				
" per-manganate, crude	lb. .60			
" phenate (phenylate, carbolate), dry....	oz. .20			
" phenol-sulphonate, see Sodium, sulphophenate (sulpho-carbolate, *U. S. Ph.*), etc....................................				
" phosphate, bi-basic (officinal), [Di-sodic ortho-Phosphate, Di-sodium Hydro-phosphate], — purified, cryst........................	lb. .25			
" " do., twice purified, cryst.........	lb. .27			
" " " " " dry..........	lb. .40			
" " " pure, granulated...........	lb. .75			
" " " chem. pure, cryst., *U. S. Ph.* and Ph. G. II.	lb. .40			
" " " " " dry	lb. .60			
" " " " " fused...........	lb. 1.25			
" " hydrochlorated (muriated), [Muriato-phosphate (Chlorhydrophosphate, Hydrochloro-phosphate) of Sodium], dry	oz. .50			
" " Meta-, see Sod., meta-phosphate .				
" phosphite...........................	oz. .60			
" phospho-lactate, see Sodium, lacto-phosphate..............................				
" " -molybdate (-molybdenate)......	oz. 1.50			
" " -wolframate (phospho-tungstate).	oz. .50			
" picro-carminate	oz. 3.00			
" plumbate ....:.....................	lb. 1.50			
" pyro-antimonate	oz. 1.00			
" pyro-borate, see Sodium, bi-borate....				
" pyro-phosphate, acid.................	lb. 2.00			
" pyro-phosphate, normal, cryst.........	lb. .90			
" " do., cryst., pure, — *U. S. Ph.* and Ph. G. II.................	lb. .94			
" " " pure, dry	lb. 1.25			
" " " fused.....................	lb. 1.50			
" " ferrated, see Iron, Sesqui-compounds: Sodio-ferric pyro-phosphate.......................				
" **quillayate**				
" rhodanide, see Sodium, sulpho-cyanate				
" ros-aniline-sulphonate..				
" rosolate............................	lb. 2.50			
" **salicylate,** pure, powder	lb. 2.65			
" " **pure, cryst.,—*U. S. Ph.*** and Ph. G. II	lb. 4.25			
" " **from Wintergreen-**(Gaultheria-)Oil ..	oz. 1.50			
" santoninate (*not* santonate!),— *U. S. Ph.*	oz. .69			
" seleniate (selenate)....................	¼oz.vls.oz.16.00			
" silicate, pure, solution [10%], —sp. gr. 1.054	lb. .50			
" " " do.,— *U. S. Ph.*,— sp. gr. 1.3-1.4 [58%]	lb. .60			
" " " cryst.	lb. 1.25			
" " " crude, lumps & ground	lb. .50			
" " " " gelatinous form ...	lb. .60			
" " " " solut'n [40–42° Bé] .	lb. .40			
N.B.—*Compare, also:* Potassium, silicate.				

☞ **When ordering, specify: "MERCK'S"!**

		Containers incl.		
Sodium, silico-fluoride.—(An innocuous surgical antiseptic, according to Thomson.) — A concentrated solution in Water contains but 0.61%		oz.	.35	
" silvate (silvinate)		oz.	1.00	
" stannate, (Mordant Salt; so-called "Preparing-salt")		lb.	.75	
" stearate		lb.	1.00	
" stibiate, Meta-, see Sodium, meta-antimonate				
" sub-sulphite, see Sodium, thio-sulphate.				
" succinate, pure, cryst.		oz.	.50	
" sulphate, (Glauber's Salt), ch. pure, cryst.		lb.	.35	
" " chem. pure, dry		lb.	.40	
" " pure, cryst.,—U. S. Ph. and Ph. G. II		lb.	.34	
" " " dry,—conforming to U.-S.-Ph. requirements		lb.	.34	
" " purified, dry		lb.	.35	
" " " cryst.		lb.	.30	
" " crude,—large crystals				
" " " —small "				
" sulphate, acid, see Sodium, bi-sulphate				
" sulphide (sulphuret), cryst., — true, — (Mono-sulphide of Sodium)		lb.	.84	
" sulphide, so-called, — (also improperly called "ter-sulphide"),—[Sodic Liver of Sulphur];—fused; and: fused, pure:—see Soda, sulphurated, etc.; etc.				
" sulphite, cryst.	(An ti- chlor.)	lb.	.26	
" " pure, dry		lb.	.50	
" " " cryst.,—U. S. Ph.		lb.	.45	
N. B. — See, also (for "Antichlor"):-Sodium, bi-sulphite; and: do., thio-sulphate.				
" " benzoated, (not a true benzoico-sulphite!),—acc. to Heckel. — [Easily soluble, powerful, innocuous antiseptic,—described as equaling the Mercury salts in force.]		oz.	.40	
" " bi-, see Sodium, bi-sulphite				
" sulpho-carbolate,—U. S. Ph.; etc.;—see Sodium, sulpho-phenate				
" " -carbonate (thio-carbonate)		lb.	.50	
" " -cyanate (thio-cyanate; rhodanide)		oz.	.30	
" " -ichthyolate (ichthyol-sulphonate), see under **Ichthyol preparations**				
" " -indigotate (sulph-indigotate; sulpho-cerulate), see Sodium, indigo-sulphate				
" " -phenate (phenol-sulphonate;—sulpho-carbolate,—U. S. Ph.), perf. white		oz.	.14	
" " " II		oz.	.13	
" " -vinate, see Sod., ethylo-sulphate				
" tannate		oz.	.30	
" tartrate, cryst.,—(NOT "Soda-Tartar"!)		lb.	.90	
" " " chem. pure		lb.	1.00	
N. B. — Tartarated (Tartarized) Soda, [Soda-Tartar], see Potassium and Sodium, tartrate.				
" tauro-cholate, (Sodium Choleate from Choleic [Tauro-cholic] Acid)		15 gr.	.75	
N. B. — Compare, also: Sodium, choleate,—Ph. G. I, - (direct from Ox Gall).				
" ter-sulphide,—improperly so called,—see Soda, sulphurated				
" thio-cyanate, see Sodium, sulpho-cyanate				

☞ When ordering, specify: "**MERCK'S**"!

		Containers incl.			
Sodium, thio-sulphate (formerly called "hypo-sulphite," or, also: "sub-sulphite")	[An Anti-chlor.]	lb. .25			
" do., chem. pure,—*Sodii hyposulphis, U. S. Ph.*		lb. .60			
N.B.—*See, also* (for "Anti-chlor"): —Sodium, bi-sulphite; and: do., sulphite.					
" tri-chlor-acetate		oz. 1.50			
" tri-chlor-phenate (tri-chlor-carbolate)		oz. .75			
" tungstate, see Sodium, wolframate					
" uranate, (Uranium Yellow;—improperly called "Yellow Oxide of Uranium")		oz. .75			
N.B.—*Compare, also*: Ammonium, uranate.					
" urate		oz. .75			
" valerianate		oz. .80			
" vanadate, pure		oz. 2.50			
" " bi-, see Sodium, bi-vanadate					
" wolframate (tungstate), crude		lb. .45			
" " purified		lb. .75			
" " pure		oz. .13			
" xanthogenate (ethylo-thio-carbonate)		oz. .30			
Sodium and Aluminium, chloride, see Aluminium and Sodium, chloride					
" and do., sulphate, see Alum, sodic					
" and **Ammonium,** oxalate		lb. 1.00			
" " " phosphate	(Microcosmic Salt.)	lb. 1.20			
" " " " ch. pure		lb. 1.35			
" " " sulphate					
" and **Copper,** chloride, see C. and S., chl.					
" and **Gold,** chloride, see Gold and Sodium, chloride, *U. S. Ph.;* and other forms and grades					
" and **Iridium,** chloride, see I. and S., chl.					
" and **Iron,** cyanide, so-called, see Sodium, ferro-cyanide					
" and do.,—*other salts,*—see under Iron, Mono-compounds; and Iron, Sesqui-compounds					
" and **Lead,** thio-sulphate ("hypo-sulphite"), see Lead and Sodium, thio-sulphate					
" and **Lithium,** salts, see Lith. and Sod.					
" and **Magnesium,** boro-citrate		oz. .40			
" " " lactate		oz. .50			
" " " phosphate		oz. .40			
" and **Mercury,** see Sodium Amalgam—(below!)					
" and **Palladium,** chloride, see Palladium and Sodium, chloride					
" and **Platinum,** double *and* triple salts, see under: Platinum double Chlorides; do. double Cyanides; and, do. *triple* Cyanides					
" and **Potassium,** carbonate, chem. pure		lb. 1.25			
" " " sulphate		lb. .75			
" " " boro-tartrate; and, tartrate (—*U.S.Ph.;* etc.);—see Pot. and Sodium, do.; and, do.					
" and **Silver,** thio-sulphate, ("hypo-sulphite"), see Silver and Sodium, thio-sulphate					
Sodium, Platinum and Potassium, cyanuret, see under Platinum triple Cyanides					
Sodium Alum, see Alum, sodic					
Sodium Amalgam		lb. 2.50			
Solanidine		15 gr. 2.25			
Solanine, pure, cryst.		15 gr. 3.00			
" hydrochlorate		15 gr. 4.00			

☞ **When ordering, specify : " MERCK'S " !**

	Containers incl.			
Soluble Citrates, so-called, see Iron, Sesqui-compounds: Ammonio-ferric citrate: brown, *U. S. Ph.*; and, green				
" **Cream** of **Tartar,** —*so-called,*—(Borax-Tartar), see Potassium and Sodium, boro-tartrate				
" do. of do., —*perfectly soluble* in Water ⎱ see do. do. do., do.,— in scales..				
" **Scales** of **Tartar** (—of *Borax-Tartar*) ⎰				
" **Glass,** (Water-Glass), see Potassium, silicate, etc.;—and: Soda, silicate, *U. S. Ph.*; etc.				
" **Indigo,** (Indigo Sulphate),—solution, —see Tinctures: Indigo				
" **Iron,** so-called, see Iron, oxide, red, saccharated				
" **Tartar,** (Tartarus tartarisatus), see Potassium, tartrate, neutral, *U.S.Ph.*; etc. N.B.—*Compare:* Soluble "*Cream,*" and "*Scales,*" of Tartar;-(*above!*).				
" do.,—**Ammoniated,**—see Potassium and Ammonium, tartrate				
Solutions (Liquores),—[*See, also:* "N. B.," at end of "Solutions"]:—				
Aluminium acetate, see Aluminium, acet., liq.				
Ammonia, aqueous, see Ammonia, Water of				
" alcoholic, see Ammonia, Spirit of				
Ammonium acetate,—Ph.G.II,—("Spiritus Mindereri")	lb. .50			
" carbonate, pyro-oleous, see Spirit, so-called, of Hartshorn,—rectified				
" succinate, ("Spiritus cornu cervi succinatus"),—sp. gr. 1.055	lb. 1.50			
" sulphide(sulphuret),—hydro-sulphuretted, — (*Hydrothion-ammonium* solution)	lb. .60			
anodyne Iron-, *Bestuscheff*"s, see Tinctures: Iron chloride,—ethereal				
Antimonious chloride, (Tri-chloride of Antimony);—[*Liquid* Butter of Antimony],—sp. gr. 1.350	lb. .35			
do. do., white, pure,—sp. gr. 1.350	lb. .50			
N.B.—*Concentrated* Butter of Antimony, see Antimony, chloride, Antimonious.				
Arsenic and Mercury Iodides,—*U. S. Ph.*;—(Solut. of Bin-iodide of Mercury and Teriodide of Arsenic),—(Donovan's Solution)				
Bamberger's Mercuro-albuminated; see Mercury, bi-chloride, albuminated, fluid.				
Chlorine,—aqueous,—see Chlorine-water				
Donovan's, see Solution, Arsenic and Mercury Iodides, *U. S. Ph.*				
Dzondi's ammoniacal, see Ammonia, Spirit				
Fehling's Test-, see under: Titrated Normal Solutions,—(*at End of Alphabetical List!*).				
Fowler's arsenical, see Solut., Potassium arsenite, *U. S. Ph.*				
Gutta-percha, — *U. S. Ph.*;—(*Traumaticin*).	lb. 3.00			
Ichthyol, see under Ichthyol preparations.				
Indigo sulphate, see Tinctures: Indigo				
Iron acetate,—sp. gr. 1.145	lb. 1.00			
" " " 1.138,	lb. .75			
" " —Ph. G. II,—sp.gr. 1.081-1.083	lb. .65			
" " —*U. S. Ph.,*— " 1.16	lb. 1.00			
" albuminate,—acc. to Dr. Friese	lb. .75			
" " " " Dr. Drees	lb. .75			
" chloride, proto- (Ferrous,)-sp.gr.1.255	lb. .35			
" " Ferric, normal, see Solution, Iron, tri-chloride				

☛ **When ordering, specify: "MERCK'S"!**

	Containers incl.		
Solutions (Liquores),—*continued:*			
Iron chloride, Ferric,-*(contin.!)*,-basic,-so-called;—see Sol., Iron oxy-chloride			
" do.,—anodyne,—see Tinctures: Iron chloride,—ethereal................			
" citrate,—*U. S. Ph.*,—sp. gr. 1.26.....			
" dialyzed,—(*a so-called Solution!*),—see Iron, dialyzed, liquid.............			
" formate,—sp. gr. 1.04..............	lb. 2.50		
" oxy-chloride, Ferric, (Basic Ferric chloride), so-called,—Ph. G. II,— [3.5% of Iron,=5% of Fe_2O_3]....	lb. .35		
" peptonized, (Peptonated Ferric Oxide), —*dialyzed;*—for *internal* use;—[3% Iron].—(Prepared from *the above.*). N. B.—*Compare, also:* Iron, peptonized, solution, glycerinated, —for *subcutaneous* injections.	lb. 1.10		
" saccharate,—*with excess of Sugar*,—see *Syrup* of Saccharate of Iron......			
" sub-sulphate,-*U.S.Ph.;*-(Sol. of Basic Ferric Sulphate), [Monsel's solution]	lb. .40		
" sulphate, Ferric, normal, (Ter-sulphate),—*U. S. Ph.* and Ph. G. I,—sp. gr. 1.32.............	lb. .50		
" " do., do., — Ph. G. II, — sp. gr. 1.428–430...........	lb. .45		
" " " " commercial.........	lb. .40		
" " " basic, see Solution, Iron, sub-sulphate, *U. S. Ph.*.			
" tri-chloride (sesqui-chloride) [Normal Ferric chloride],—sp. gr. 1.500	lb. .85		
" " —sp. gr. 1.480.............	lb. .75		
" " " 1.405,—*U. S. Ph.*.....	lb. .65		
" " " 1.28,—Ph. G. II.....	lb. .50		
Lead acetate, basic, (sub-acetate),—[so-called Goulard's Extract; Vinegar of Lead—Acetum plumbi (Saturni)],—*Liquor plumbi subacetatis, U. S. Ph.*.................	lb. .30		
Lime,—*U. S. Ph.*,—(Lime-water—Aqua Calcariæ).....................	lb. .25		
Mercuric nitrate, (Mercury Per-nitrate),— sp. gr. 1.180...........	lb. 1.10		
" " —sp. gr. 2.10,—*U. S. Ph.*...	lb. 2.00		
" " " 1.67.............	lb. 1.60		
Mercury bi-chloride, albuminated,—according to Bamberger,—see Mercury, bi-chloride, albuminated, fluid............			
Monsel's, see Solution, Iron sub-sulphate, *U. S. Ph.*			
pancreatic,—prepared directly from the fresh pancreas;—(*not* Glycerolate of Pancreatin!—*which see also*, under: Pancreatin,—solution in Glycerin.).............	lb. 1.50		
Potassa, caustic,—sp.gr.1.340 ⎰ [34% Potass.	lb. .30		
" " pure,- " 1.340 ⎱ Hydr.-KHO]	lb. .75		
" " " " 1.142,- Ph. G. II, - [15% of KHO].	lb. .40		
Potassium acetate,—Ph. G. H..........	lb. .75		
" arsenite,—*U. S. Ph.;*—(Fowler's Arsenical solution)................			
" silicate, (Liquid Glass), see under: Potassium, silicate	lb. .40		
Soda, caustic, — sp.gr.1.340. ⎰ [31% Sodium	lb. .30		
" " pure, " 1.340. ⎱ Hydr.-NaHO]	lb. .75		
" " " " 1.159–163,-Ph.G. II, —[abt.15% NaHO]	lb. .40		
" " —sp. gr. 1.34,—[37° Bé];—free from Nitrogen.—[For determining Nitrogen in analyses.]	lb. .35		

☛ **When ordering, specify : "MERCK'S"!**

	Containers incl.	
Solutions (Liquores),—*continued:*		
Sodium ethylate, (Liquor Sodii ethylatis, Ph. Brit.), see Sodium, ethylate, liquid		
" hypo-chlorite	lb. .35	
" silicate, (Liquid Glass), - *U. S. Ph.;* and other grades;—see under: Sodium, silicate		
N. B.—*Many other Solutions,* see under the names of the various Metallic salts, etc. —*Compare, also:* TINCTURES, etc.; and, SYRUP, etc.		
Solutions, Test-, (*Indicator-, titrated normal, and pharmacopœial volumetric Solutions*),— for qualitative and quantitative analyses,— see at End of List.		
Sorbin (Sorbinose).....................	15 gr. 1.50	
Sorbit (Sorbitol).......................		
Sozo-iodole (Di-iod-para-phenol-sulphonate of Sodium),—readily soluble................	oz. 1.75	
☞ N.B.—*The analogous* salts of **Potassium, Ammonium, Barium, Lead, Mercury, Silver,** and **Zinc,** are also made.		
Sparteine Merck:		
pure Alkaloid,-syrupy consistency.-(Narcotic.)	15 gr. .50	
hydrochlorate, cryst.	15 gr. .50	
hydro-iodate (hydriodate), **cryst.**,—readily soluble in 5 parts of Water..............	15 gr. .50	
sulphate, cryst.	15 gr. .30	
Specimen Collections:		
Alkaloids, Glucosides, etc.		
All the Opium constituents	See at End of List.	
Metals		
Physiological Preparations		
Spigeline.—The highly toxic active principle of Maryland Pink—*Spigelia marilandica.*— (Anthelmintic; specially in ascarides!).....		
N.B.—*See, also:*—Fluid Extracts: Spigelia.		
Spirit, Angelica,—compound.............	lb. .85	
" aromatic,—Ph. Neerl.	lb. 1.00	
" Balm (Lemon - balm — Melissa),—compound; ["Eau des Carmes"].	lb. 1.00	
" " —simple, concentrated.........	lb. 1.50	
" Cochlearia (Scurvy-grass, Spoonwort), —Ph. G. II,—from the fresh herb....	lb. 1.00	
" Elder-flowers, see Spirit, Sambucus....		
" formic, (Spirit of Ants—Spiritus Formicarum),—*true,*—prep. from ants	lb. 1.00	
" " —Ph. G. II,-prep. fr. Formic Acid.	lb. .90	
" Mastic (Mastix),—compound; (Spiritus matricalis—Mother-spirit)..........	lb. 1.50	
" Melissa: compound; and simple;—see Spirit, Balm		
" —*so-called,*— Mindererus's, see Solutions: Ammonium acetate		
" Mother-, see Spirit, Mastic,—compound		
" pyro-acetic,—so-called,—see Acetone..		
" pyro-ligneous (pyro-xylic), see Alcohol, methylic		
" Raspberry;— for preparing Aqua Rubi idæi	lb. 1.50	
" Sambucus (Elder-flowers).............	lb. 1.50	
" Scurvy - grass (Spoonwort), see Spirit, Cochlearia......................		
" Wood-, see Alcohol, methylic.........		
Spirit of Ammonia, Dzondi's, see Ammonia, Spirit of		
" " " —aromatic..............	lb. 1.00	
" of Ants, see Spirit, formic............		
" —*so-called,*—fuming, of Libavius; see Tin, tetra-chloride		

☞ When ordering, specify: "**MERCK'S**"!

	Containers incl.		
Spirit — *so-called* — of Hartshorn, — *rectified;* (Spiritus Cornu Cervi rectificatus; Liquor Ammonii carbonici pyro-oleosi—Solution of Pyro-oleous Ammonium Carbonate)................	lb. .60		
" —*so-called*—of Hartshorn, —*succinated;* see Solutions: Ammonium succinate			
" of Iron Chloride, —etherized; see Tinctures: Iron chloride, —ethereal......			
" of Muriatic Ether; (*Sweet Spirit of Salt*), [Hydrochlorated Alcohol], — sp. gr. 0.840................................	lb. 1.25		
" of Nitrous Ether; (*Sweet Spirit of Nitre*), —*U. S. Ph.*..............................			
Spiritus æthereus martiatus, (Spir. Ferri chlorati æthereus), see Tinctures: Iron chloride, —ethereal..................			
" **Ammoniaci caustici Dzondii,** see Ammonia, Spirit of................			
" **Cornu Cervi rectificatus,** see Spirit, so-called, of Hartshorn, —rectified			
" " " **succinatus,** see Solutions: Ammonium, succinate..			
" **fumans Libavii,** see Tin, tetra-chloride.............................			
Spiritus, other than above, see: Spirit, etc....			
Spodium purificatum; et, purum;—see Charcoal, animal, purified, *U. S. Ph.*; and, pure			
Sponge, burnt, (Spongia usta [tosta]), see *Charcoal,* Sponge-...			
" compressed, (Spongiæ pressæ), — tied with twine...................	oz. .75		
" " in layers, —without twine	oz. 1.50		
Sponge-tent (*Waxed Sponge* — Spongiæ ceratæ)....................................	oz. .70		
Squill (Scilla) preparations:			
Scilli-picrin Merck	15 gr. .35		
Scillitin	15 gr. .75		
Scilli-toxin (Scillain).....................	15 gr. 2.00		
Stanni pulvis, see Tin, metallic, pure, powder...............................			
Stannic Precipitate of Gold, see Gold, Tin-precipitate of......................			
Stannum, and compounds, see Tin, etc....			
Staphisagrine...........................	15 gr. 1.00		
Starch (Amidin, Fecula), iodized,—(Amylum iodatum, *U. S. Ph.*);—["Iodide of Starch"], —soluble	oz. .34		
" of Inula (-of *Elecampane;* -of *Alant*-root),—[Alant-starch ; Alantin ; Dahlin], — see Inulin			
Starch-sugar, chem. pure, anhydrous, see **Grape-sugar,** etc.............................			
Steel Pellets, so-called, see Iron, Mono-compounds: Potassio - Ferrous tartrate, in globules....................			
Stibium, and compounds, see Antimony, etc. (—"*Stibiated*—" etc., see "Antimoniated—" etc.)....................................			
Stilbene (Symmetric Di-phenyl-ethylene) [Toluylene]............................	15 gr. 1.00		
Stone, divine .... } so-called, see Copper, " **ophthalmic,** } aluminated......... " **infernal,** see Silver, nitrate, cryst.; and, molded;—*U. S. Ph.*; and, grey......			
Strontium, metallic,—from Amalgam	15 gr. 5.00		
" " —by electrolysis	15 gr. 10.00		
" acetate.....................	lb. 2.50		
" bromate......................	oz. 1.00		

☞ **When ordering, specify : "MERCK'S"!**

	Containers incl.			
Strontium, bromide	oz. .50			
" carbonate, pure, perf. white	lb. .60			
" chlorate	lb. 1.85			
" chloride, chem. pure, cryst.	lb. 1.25			
" " cryst.	lb. .75			
" " dry	lb. 1.50			
' chromate	lb. 2.25			
" fluoride. — (An inhalant in laryngeal phthisis.)	lb. 2.25			
" formate	oz. .50			
" hypo-sulphate	oz. .75			
" hypo-sulphite, see Strontium, thio-sulphate				
' iodide	oz. 1.00			
" nitrate, pure, anhydrous, cryst.	lb. 1.00			
" " dry	lb. .25			
' oxalate	lb. 1.30			
" oxide, caustic, cryst.	lb. 1.50			
" " anhydrous	lb. 2.00			
" phosphate	lb. 1.50			
" sulphate, precipitated	lb. 1.00			
" sulphide (sulphuret)	lb. 1.50			
" thio-sulphate (formerly called "hyposulphite")	oz. .75			
Strontium and Platinum, cyanide, see under Platinum double Cyanides				
" **and Potassium,** chlorate	lb. 2.50			
Strophanthin Merck, chem. pure, cryst.; — from Strophanthus hispidus, an African arrow-poison. — (Preferred to Digitalin, — as a heart-tonic.)	grain .50			
Strychnine (Strychnia), pure, cryst.,—*U. S. Ph.*	⅛ oz.vls.oz. 2.00			
" **pure, precipitated**	⅛ oz.vls.oz. 1.95			
" **acetate**	⅛ oz.vls.oz. 2.00			
" **arseniate** (arsenate)	⅛ oz.vls.oz. 3.50			
" **arsenite**	⅛ oz.vls.oz. 4.00			
" **camphorate**	⅛ oz.vls.oz. 6.00			
' **citrate**	⅛ oz.vls.oz. 6.00			
" **ferri-citrate,**—*Ferri et Strychninæ citras, U. S. Ph.*	oz. 1.00			
" **hydrobromate**	⅛ oz.vls.oz. 6.00			
" **hydrochlorate**	⅛ oz.vls.oz. 2.00			
" **hydro-iodate** (hydriodate)	⅛ oz.vls.oz. 6.00			
" " —**with Iodide of Zinc**	⅛ oz.vls.oz. 4.00			
" **hypo-phosphite**	⅛ oz.vls.oz. 3.50			
" **lactate**	⅛ oz.vls.oz. 4.00			
" **nitrate,** cryst.	⅛ oz.vls.oz. 2.00			
" **phosphate**	⅛ oz.vls.oz. 3.00			
" **saccharinate** (*not* saccharate!) *True salts of Strychnine and Saccharin—which latter see!*				
" " **bi-**...*U. S. Ph.*				
" **sulphate,**—*U. S. Ph.*	⅛ oz.vls.oz. 2.00			
" **sulpho - carbolate** (phenol - sulphonate, sulpho-phenate)	⅛ oz.vls.oz. 5.00			
Strychnine and Zinc-Oxide, hydriodate, see Str., hydro-iodate,—with Iodide of Zinc				
Strychnine with Ferri-citrate of Quinine	⅛ oz.vls.oz. 3.00			
Strychnine, Methyl-, etc., see Methyl-Strychnine, etc.				
Styracin, cryst., white, (Cinnamate of Cinnyl [Styryl]), [Cinnamylo-cinnamic Ether]	oz. 5.00			
Styrol (Styrolene; Cinnamene, Cinnamol), chem. pure	oz. 2.50			
Styrone (Cinnyl Alcohol; Cinnamic [Styrylic] Alcohol), liquid	oz. 2.00			
" cryst.	oz. 5.00			
Suberin	oz. .65			
Sublimate, corrosive, see Mercury, bichloride, *U. S. Ph.;* etc.				
Succus. Succi, etc., see Juice, Juices, etc.				

☞ **When ordering, specify: "MERCK'S"!**

	Containers incl.		

Sugar, ferruginated, (*Iron-Sugar*), see Iron, oxide, red, saccharated ....●....
   N.B.—*Compare, also:*
     Iron, albuminate .... ⎫
       " carbonate—(*U. S. Ph.;* etc.)— ⎪ *saccharated*
       " iodide—(*U. S. Ph.*) - .... ⎬
       " peptonized .... ⎪
       " sulphate, Ferrous .... ⎪
       " Mono-compounds: Mangano- Ferrous carbonate .... ⎭
Sugar, Grape-, ⎱ (Dextrose, Dextro-glucose;
  " Starch-, ⎰ Glucose,)—see Grape-sugar,
    chem. pure, anhydrous, etc. ....
  " Fruit-, (Levulose), see Fruit-sugar, I..
  " inverted, see Fruit-sugar, commercial
  " Madagascar, see Melampyrit........
  " Milk-, (Lactose, Lactin), see Milk-sugar
  " of Acorns, see Quercit ....
  " of Manna, see Mannit ....
  " of Meat, see Inosit ....
Sugar—*so-called*—of Lead, see Lead, acetate, normal, *U. S. Ph.*...................
Sulfur, etc., = Sulphur, etc..............
Sulpho-phenol (Sulpho-carbol), para- and ortho-, —*mixed*, —see Acid, sulpho-carbolic ....
  " ortho-, *pure*, -33⅓%, solution, -see Aseptol
Sulpho-urea (Sulph-urea) [Sulpho-carbamide] ....   oz. 3.00
Sulphonal (Di-ethyl-sulphon-di-methyl-methane) [= (C H₃)₂.C.(C₂H₅.SO₂)₂].—Crystals, soluble in 500 parts Water of 15° C [59 F]; in 65 of Absolute Alcohol, or in 110 of 50-% Alc., at same temperature.—(Reported to be a non-narcotic hypnotic, without heart-effects.) ....   oz. 2.25
Sulphur, sublimed, (Flowers of Sulphur),— *Sulphur sublimatum, U. S. Ph.* ....
  " do., washed (purified), [Washed Flowers of Sulphur],—*Sulphur lotum, U. S. Ph.*
  " precipitated, (Milk [Magistery] of Sulphur—Lac Sulphuris), pure,— *Sulphur præcipitatum, U. S. Ph.*   lb. .35
  "   " commercial ....   lb. .20
  " chem. pure, cryst. ....   lb. 1.00
  " bromide ....   oz. 1.00
  " chloride.........................   oz. .50
  "   " camphorated..................   oz. .75
  " di-oxide, hydrated, —solution,—see Acid, sulphurous,—*U. S. Ph.;* etc......
  " —*so-called*,—golden,—(Sb₂S₅);—see Antimony, sulphide, golden ....
  " iodide,—*U. S. Ph.* ....   oz. .50
  " tri-oxide, see Acid, sulphuric, anhydrous
  "   " mono-hydrated, see Acid, sulphuric, chem. pure, *U. S. Ph.*...
Sulphur stibiatum aurantiacum,(*Sulphur auratum Antimonii*), — [*not:* "Sulphurated Antimony," U. S. Ph.;—*but: Penta-sulphide* of Ant.!];—see Antimony, sulphide, golden
Sulphur, —so-called "Alcohol" of,—see Carbon, bi-sulphide ....
  " Balsam of, see Oils, divers: sulphurated Linseed-......................
  " do. do., terebinthinated, see Oils, divers: sulphurated Linseed-, terebinthinated
  " Flowers of, see Sulphur, sublimed, *U. S. Ph.* ....
  " do. do., washed, see Sulphur, sublimed, washed, *U. S. Ph.* ....

☞ **When ordering, specify: "MERCK'S"!**

	Containers incl.
Sulphur, Liver of, (*Potassic Liver of Sulphur*), see Potassa, sulphurated, *U. S. Ph.*; and other grades.............	
" do. do., **calcic**, see Lime, sulphurated, *U. S. Ph*..................	
" " " do., **antimoniated** (*stibiated*), see Lime, antimonio-sulphurated...............	
" " " **sodic**, see Soda, sulphurated, etc.....................	
" **Milk** (*Magistery*) of, see Sulphur, precipitated, *U. S. Ph.*; etc............	
Syringin............................	15 gr. 2.50
Syrup, Buckthorn (Common [purging] Buckthorn), — [Syrupus Spinæ cervinæ; Syr. Rhamni catharticæ (*cathartici*)].	lb. .60
" Cherry, (Syrupus Cerasorum)........	lb. .75
" Mulberry, (Syrupus Mororum).......	lb. .60
" Papaw (Carica Papaya).—[100 grammes dissolve 250 grammes of meat.].....	oz. 1.00
" Poppies (Poppy-capsules), [Syrupus Diacodii (Papaveris; capitum Papaveris)]	
" Raspberry, (Syrupus Rubi idæi).......	lb. .50
" of Saccharate of Iron, (Syrup of Saccharated Ferric Oxide; Syrup of Soluble Saccharated Oxide of Iron)	lb. .75
" Violets, (Syrupus Violarum)..........	lb. 1.00

☞ **When ordering, specify: "MERCK'S"!**

	Containers incl.			

☞ When ordering, specify: "MERCK'S"!

		Containers incl.			
Tannin (Tannic Acid), very light, chem. pure, clearly soluble, — *U.S.Ph.* and Ph. G. II .... } *consistency foamy*	oz.	.30			
" very light, pure	oz.	.28			
" commercial, powder or granulated, I .... } *soluble in Water and in Alcohol*	lb.	2.00			
" " powder or granulated, II	lb.	1.95			
" " powder, III	lb.	1.90			
" " " IV	lb.	1.85			
" powder,—Ph. G. II,—perfectly white..	oz.	.25			
" odorless and soluble	oz.	.35			
" *in sticks*	oz.	.50			
Tannin Albuminate	oz.	.50			
Tantalum, metallic, pure	15 gr.	7.50			
" pent-oxide, (Tantalic Oxide), hydrated, —from Tantalic Chloride;—see Acid, tantalic					
Tar (Pix) of **Birch**, see Oils, divers: Birch; empyreumatic					
" of **Juniper** (Juniper-wood), see Oils, divers: Cade					
" of **Lignite**, see Oils, divers: Lignite..					
Tartar, chem. pure, see Potassium, bi-tartrate, *U. S. Ph.;* etc.					
" Cream of, } see Potassium, bi-tartrate, " Crystals of, } *U. S. Ph.;* etc.; etc.... N.B.—*Compare, also:* Tartar, *Soluble* Cream of, ("so-called"; and, "*perfectly* soluble"),—below!					
" purified; and, pure; (Crystals of Tartar; Cream of Tartar);—see Potassium, bi-tartrate, etc., etc.					
Tartar, ammoniated, soluble, see Potassium and Ammonium, tartrate					
" **ammonio-ferric**, (*Ammoniacal Iron-Tartar*), see Iron, Sesqui-compounds: Ammonio-Ferric tartrate, *U. S. Ph.* ...					
" **antimoniated**, (Tartarus stibiatus), [Tartar Emetic], see Antimony and Potassium, tartrate, *U. S. Ph.;* and other grades					
" **Borax-,** (Tartarus boraxatus), [*so-called* "*Soluble* Cream of Tartar"], see Potassium and Sodium, boro-tartrate...					
" **do.-,** *perfectly soluble* in Water!—see do. do. do., do.,—*in scales*					
" **essential Salt** of, see Acid, tartaric... N.B.—*Compare:* Tartar, *Salt* of,—(below)!					
" **ferrated,** } see Iron, *Mono*-compounds: " **Iron-** ... } Potassio-*Ferrous* tartrate. N.B.—*Compare:* Tartarated (Tartarized) Iron,—[below]!					
" **ferrid-ammoniacal,** } *see Iron, Sesqui-compounds: Ammonio-Ferric tartrate, U.S.Ph.* " **Iron-, ammoniacal,** }					
" **Salt** of, see Potassium, carbonate, *pure*. N. B.—*Compare:* Tartar, *essential* Salt of,—(above)!					
" **Soda-,** see Potassium and Sodium, tartrate, *U. S. Ph.;* etc.					
" **soluble,** (Tartarus tartarisatus), see Potassium, tartrate, neutral					
" " **ammoniated,** see Potassium and Ammonium, tartrate					
" **soluble Cream** of,—*so-called*,—(Borax-Tartar), — see Potassium and Sodium, boro-tartrate					
" " " **do. do.** ... } *—perfectly soluble in Water!* } *—see do. do. do., do.,—in scales* " " " **Scales** of, }					

☞ When ordering, specify: " **MERCK'S !** "

	Containers incl.			
artar,—*(continued !)*,—**tartarized** *(tartarated)*, [Soluble Tartar], see Potassium, tartrate, neutral..................				
" **vitriolated**, see Potassium, sulphate..				
artar **Emetic**............ } see Antimony and Potassium, tartrate, *U. S. Ph.*; and other grades.				
Tartarus stibiatus, (Antimoniated Tartar).........				
Tartarated *(Tartarized)* **Antimony**............				
" **Iron**, see Iron, *Sesqui*-compounds : Potassio-*Ferric* tartrate, *U. S. Ph.*......				
N.B.—*Compare:* Tartar, ferrated, (Iron-Tartar),—[above]!				
" **Soda,** (Soda-Tartar), } see Potassium and Sodium, tartrate, *U. S. Ph.*; etc.............				
Tartarus natronatus...				
" **boraxatus,** (Borax-Tartar), [Cremor Tartari *quasi solubilis!*], see Potassium and Sodium, boro-tartrate				
" do.,—*plane solubilis!*—see do. do. do., do.,—*in scales*				
" **tartarisatus,** (Soluble Tartar), see Potassium, tartrate, neutral				
Taurine (Amido-ethyl-sulphonic Acid).....	15 gr. 2.50			
Tellurium, pure......................	15 gr. 1.00			
" di-oxide, (Tellurous oxide), hydrated, —[Tellurous Hydroxide];—see Acid, tellurous.......................				
" tri-oxide, (Telluric oxide), tri-hydrated, —[Di-hydrated Telluric Hydroxide]; —see Acid, telluric, di-hydrated.....				
Terebene,—optically inactive.................	lb. 1.00			
" Dr. Bond's,—in original bottles.......	each .75			
Terpenes,—optically active,—hydrochlorates of, see Turpentine-oil, etc.; etc...				
Terpin Hydrate, cryst.—(Ter-hydrate of *optically inactive* Terpenes).—[Succedaneum for Turpentine-oil.]	oz. .35			
Terpinol, liquid........................	oz. .65			
Terra foliata Tartari, see Potassium, acetate, *U. S. Ph.;* and other grades and forms				
Terra foliata Tartari crystallisata, see Sodium, acetate, *U. S. Ph.;* and other kinds				
Test-papers, see Paper, etc...............				
Test-solutions (*Indicator-, titrated normal,* and *pharmacopœial volumetric Solutions*),—for qualitative and quantitative analyses,—see at End of List.				
Tetr-iod-pyrrole, see Iodole............				
Thalline (Tetra-hydro-para-chin-[quin-]anisol),—[Methyl-ether of Tetra-hydro-para-oxy-quinoline],—salicylate.....	oz. 2.50			
" sulphate	oz. 2.50			
" tannate	oz. 1.75			
" tartrate	oz. 2.25			
Thallium, metallic.....................	15 gr. .30			
" oxide	15 gr. .50			
Thallium-salts:—Acetate; bromide; carbonate; chloride; sesqui-chloride; iodide; nitrate; sulphate................[*each:*—	15 gr. .50			
Thebaine, pure.....................	15 gr. .65			
" hydrochlorate	15 gr. .65			
" tartrate, acid	15 gr. .65			
Theine, see Caffeine....................				
Theobroma, Oil of, see Butter, Cacao-....				
Theobromine	15 gr. 1.25			
" hydrochlorate, cryst...............	15 gr. 1.25			
Thermifugin (Methyl-tri-hydro-oxy-quinoline-carbonate of Sodium);—[*formula of the Acid:* see under Acids!].—(An antipyretic, discovered by Prof. Demme, of Berne.)....				

☞ When ordering, specify : "**MERCK'S**"!

	Containers incl.			
Thio-alcohol, ethylic, see Mercaptan, ethylic				
Thorium, metallic	15 gr. 20.00			
" sulphate	15 gr. 3.50			
Thridace, see Lactucarium, Gallic				
Thymol, cryst., — *U. S. Ph.*,—(Thymic Acid; *Thyme-camphor*)	oz. .49			
Thymol-Mercury, acetate, (Thymol-acetate of Mercury), see Mercur-Thymol, acetate				
Tin (Stannum), double salts of, see "Tin and —" (below!)				
" metallic, pure, in sticks	lb. 1.00			
" " " granulated	lb. 1.00			
" " " precipitated	lb. 1.50			
" " " powder, (Stanni pulvis)	lb. 1.50			
" " " filings	lb. 1.00			
" ammonio-chloride, see Tin and Ammonium, chloride				
" bi-chloride, *fuming,–so-called,*–(Libavius's "Spirit"), see Tin, tetra-chloride				
" " *cryst.*, white, —*so-called,*—see Tin and Sodium, chloride				
" " *true*, see Tin, chloride				
" bi-sulphide (bi-sulphuret)	oz. .30			
" chloride (di-chloride — *true bi-chloride;* —formerly called "proto-chloride"), [Stannous chloride], — pure; ± (*Anhydrous form* of the so-called "Tin-salt")	lb. .70			
" iodide	oz. 1.00			
" oxalate	lb. 2.50			
" oxide, white, (per-oxide, di-oxide), (Stannic oxide; Anhydrous Stannic Acid)	lb. .90			
" " do., pure, (Flowers of Tin — Flores Jovis [Stanni])	lb. 1.00			
" oxide, grey, (Tin Ash—Cinis Jovis [Stanni]).—[Used in the arts as so-called *Putty-powder* (Polishing-powder).]	lb. .70			
" oxide, black, (prot-oxide, mon-oxide, [Stannous oxide], pure	lb. 1.50			
" phosphide (phosphuret), mono-	oz. .75			
" sulphate, Stannous [Protoxide salt]	oz. .25			
" sulphide (sulphuret), cryst.	oz. .25			
" tannate	oz. .65			
" tartrate	oz. .45			
" tetra-chloride, (so-called "Fuming Bi-chloride"; Spiritus fumans Libavii); [Stannic chloride; *Anhydrous* Butter of Tin]	oz. .40			
Tin and Ammonium, chloride, (Ammonio-stannic chloride; Chloro-stannate of Ammonium), [Pink Salt; Dyers' Salt]	lb. .65			
" and **Mercury** and **Zinc,** *Amalgam,* see Zinc and Tin, Amalgam				
" and **Sodium,** chloride, (*so-called* "White Crystallized Tin Bi-chloride")	lb. .65			
Tin and Zinc, Amalgam, see Zinc and Tin, Amalgam				
Tin-precipitate of **Gold,** see Gold, Tin-precipitate of				
Tin Ash, see Tin, oxide, grey				
" **Butter,** *anhydr.*, see Tin, tetra-chloride				
" **Flowers,** see Tin, oxide, white, pure				
" **Powder,** see Tin, metallic, pure, powder				
" **Salt,** so-called,—*anhydrous,*—see Tin, chloride				
Tinctures:				
Aconite: root (tuber),—Ph. G. II	lb. 1.25			
Actæa, see Tincture, Cimicifuga				
Adonis vernalis, (Bird's Eye; *False* Hellebore): herb	lb. 1.50			

☞ **When ordering, specify: "MERCK'S"!**

MERCK'S INDEX.

	Containers incl.			
Tinctures,—*continued:*				
Ants,—(Tinctura Formicarum),—Ph. G. I..	lb. 1.25			
Arbor vitæ, see Tincture, Thuja				
Arnica: flowers..........................	lb. 1.25			
Arnica: fresh herb......................	lb. 1.50			
arsenical, Fowler's, see Solutions: Potassium arsenite, *U. S. Ph.*				
Belladonna: fresh leaves,—Ph. G. I	lb. 1.25			
Bestuscheff's, see Tincture, Iron chloride, —ethereal				
Bryony,—from the juice of the fresh root..	lb. 1.25			
Cactus grandiflorus, (Night - blooming Cereus)................................				
Caladium seguinum, see Tinct., Dumb-cane				
Cannabis, Indian,—Ph. G. II,—(Alcoholic ·5-% solution of Extract of Indian Hemp).	lb. 1.25			
Capparis: seed, see Tincture, Simulo......				
Carduus marianus, (Mary-Thistle), — Ph. G. I....................................				
Cascara sagrada, (Chittem-bark)...........	lb. 1.50			
Celandine: herb,—according to Rademacher	lb. 1.50			
Chamomile, German, (Matricaria chamomilla): dried flower-heads,—Ph. G. I ...				
Cimicifuga (Actæa): root	lb. 1.25			
Cochineal,—Ph. G. II.....................	lb. 1.25			
Condurango (Mataperro): bark...........	lb. 2.00			
Conium: herb............................	lb. 1.25			
Convallaria: entire plant	lb. 1.50			
Copper acetate,—acc. to Rademacher	lb. 1.50			
Coto-bark	lb. 1.50			
Damiana: leaves.........................	lb. 1.75			
Digitalis: dry leaves,—Ph. G. II...........	lb. 1.25			
Drosera rotundifolia, (Rorella), [Roundleaved Sundew]: dry herb,—Ph. G. I...				
Dumb-cane (Caladium seguinum): root....	lb. 1.50			
Eucalyptus: leaves	lb. 1.25			
Garcinia, see Tincture, Mangosteen				
Gelsemium: root	lb. 1.25			
Geranium: root, (Cranesbill-root).........	lb. 1.50			
Guaco: herb	lb. 1.50			
Hamamelis: bark	lb. 1.25			
Hellebore, Green, *American*, see Tincture, Veratrum, Green.................				
" White, *European*, see Tincture, Veratrum, White..................				
" *False*, see Tincture, Adonis vernalis..				
Hydrastis: root...........................	lb. 1.25			
Hyoscyamus: fresh herb..................	lb. 1.25			
Indigo,—(Solution of "Soluble Indigo" [-of Indigo Sulphate])	lb. 1.25			
Iodine; dark,—Ph. G. II,—(10-% alcoholic solution)	lb. 1.50			
" decolorized,—Ph. G. I.............	lb. 1.75			
" -- Ph. Brit.	lb. 1.60			
Iron acetate,—ethereal,—Ph. G. II........	lb. 1.25			
" " —acc. to Rademacher..........	lb. 1.25			
Iron chloride, — ethereal; — (Bestuscheff's tonico-nervine Tincture), [Etherized Spirit of Iron Chloride,—Liquor anodynus martiatus]..................................	lb. 1.50			
Lacmus (Chemically Pure Litmus).—[Indicator Solution.]	lb. 1.50			
N. B.—*See, also*, under: Indicator Solutions (*Test-solutions*), at End of List.				
Lactuca virosa, (Acrid Lettuce): fresh flowering herb,—Ph. G. I				
Lippia mexicana: herb...................	lb. 1.75			
Mangosteen (Garcinia): fruit rind,—ethereal	lb. 1.75			
Matricaria, see Tincture, Chamomile, German....................................				

☞ **When ordering, specify : "MERCK'S"!**

	Containers incl.			
Tinctures,—*continued:*				
Musk,—Ph. G. II....................	oz. 1.50			
Nutgalls,—Ph. G. II..................				
Nux vomica,-(Tinctura Strychni),-Ph.G.II.	lb. 1.00			
Opium; simple,—Ph. G. II,—(Laudanum)	lb. 1.50			
" saffronated,—(Tinctura Opii crocata), —Ph. G. II;—[Sydenham's Laudanum; so-called "Wine of Opium"].				
Poison-oak, see Tincture, Rhus toxicodendron.................................				
Pulsatilla: fresh herb...................	lb. 1.25			
Quebracho blanco: bark...............	lb. 1.35			
do. do.; do.,—*acc. to Penzoldt*,—see Extracts: Quebracho blanco,—acc. to Penzoldt,—*liquid*				
Quebracho colorado: wood...............	lb. 1.25			
Rennet, see Rennet Wine...............				
Rhus toxicodendron, (Poison-oak): leaves..	lb. 1.25			
Simulo (Capparis-seed).—[A nervine, according to Christy.]				
Spilanthes; compound,—(also called: "*Paraguay roux*").........................	lb. 1.50			
Staphisagria: seed	lb. 1.25			
Stramonium				
Strophanthus: seed,—strength, 1:20........	lb. 1.75			
" " " 1:10........	lb. 2.50			
Strychnos-seed,—Ph. G. II,—see Tincture, Nux vomica...........................				
Tayuya-root, from Trianosperma ficifolia,— strength, 1:9...........................	lb. 2.50			
Thuja (Arbor vitæ): leaves................	lb. 1.35			
Vanilla: pod............................	lb. 3.00			
Veratrum, Green, (*American* Green Hellebore; Indian Poke): rhizome...........	lb. 1.25			
Veratrum, White, (*European* White Hellebore): rhizome,—Ph. G. II............				
Viburnum prunifolium, (Black Haw): bark.	lb. 1.75			
Titanium, metallic......................	15 gr. 2.50			
" chloride.....................	15 gr. .30			
" di-oxide, di-hydrated, (Titanic Hydroxide), see Acid, titanic, Ortho-......				
Titanium and **Potassium**, fluoride.:......	oz. 3.00			
Titrated Normal Solutions, (Test-solutions), see at End of List.				
Toluene (Toluol) [Methyl-benzene; Phenylmethane], pure,—sp. gr. 0.877; m.-p. 110–112°C [230–233.6 F]............	lb. .65			
" di-Amido-, see Tolylene-di-amine......				
" mono-chlorated, see Mono-chlor-toluene				
Toluidine, (Amido-toluene [-toluol]; Tolylamine), **ortho-,** commercial	oz. .25			
" do., chem. pure......................	oz. .50			
" **para-,** commercial...................	oz. .25			
" " chem. pure......................	oz. .50			
" " sulphate........................	oz. 1.50			
Toluylene, see Stilbene..................				
Tolyl-amine, see Toluidine				
Tolylene-di-amine (Di-amido-toluene [-toluol])—[sometimes mis-called: *Toluy*-*lene*-di-amine]	oz. 3.50			
Tonka-bean Camphor, see Cumarin......				
Traumaticin, see Solutions: Gutta-percha, *U. S. Ph.*............................				
Tri-butyrin, see Butyrin				
Tri-chlor-methyl, sulphite, (Tri-chlormethyl-sulphonic Acid)................	oz. 6.00			
Tri-chlor-phenol, cryst.,—m.-p. 65°C [149F]	oz. .45			
Tri-ethyl-amine	oz. 6.00			
" hydrochlorate.....................	oz. 5.00			

☞ When ordering, specify: "MERCK'S"!

	Containers incl.			
Tri-methyl-amine (often *erroneously* prescribed or ordered by the name of "Propyl-amine"),—10-% solution, aqueous	oz. .35			
" hydrochlorate	½ oz.vls.oz. 4.00			
" sulphate	15 gr. .40			
Tri-methyl-carbinol (Tertiary Butylic Alcohol),—deliquescent crystals; melt.-point 25° C [77 F]; boil.-pt. abt. 85° C [185 F]	oz. 3.00			
Tri-oxy-benzo-phenone, see Salicyl-Resorcin-ketone				
Tri-stearin	oz. .75			
Tropeolin (Tropæolin) 00 ......(orange W.)	oz. .50			
" ....000 No. 1 (" I.)	oz. .40			
" ....000 No. 2 (" II.)	oz. .45			
N.B.—*Tropeolin* "00" is used as an Indicator in *Soda-testing;* "000 No. 2" as an Indicator for *Acids*.				
Tropine, pure	15 gr. 1.50			
" sulphate	15 gr. 1.50			
Trypsin.—The Albumen-solving constituent of Pancreatin	oz. 4.00			
Tungsten, etc., see Wolfram, etc.				
Turmeric Paper, see under Paper				
" Yellow, see Curcumin				
Turpentine-oil, mono-hydrochlorate, solid, white, (so-called "Artificial Camphor")	oz. .65			
" di-hydrochlorate, (so-called "Lemon Camphor")	oz. 1.00			
Turpeth, ammoniacal, see Mercury and Ammonium, sulphate				
" nitric, see Mercury, nitrate, Mercurous, basic				
Turpeth Mineral, see Mercury, sulphate, Mercuric, basic,—*U. S. Ph*				
Turpethin, see Resins: Turpeth-root				
Tyrosin	15 gr. 2.00			

☞ When ordering, specify: "MERCK'S"!

	Containers incl.			
Unguentum, see Ointment				
Uranin.—A coal-tar-dye generator	oz. .75			
Uranium, metallic, fused	15 gr. 3.00			
" acetate, pure.—(For analyses.)	oz. .80			
" bromate				
" bromide	oz. 1.50			
" chloride	oz. .80			
" nitrate, cryst, ch. pure.--(For analyses.)	oz. .90			
" oxalate, cryst.	oz. 1.50			
" oxide, yellow,—so-called;—("Uranium Yellow"):—see *Sodium*, uranate				
" oxide, hydrated,—so-called;—(sometimes also called "Uranium Yellow"):—see *Ammonium*, uranate				
" oxide, black—(*principally:* Uranoso-uranic Oxide),—pure	oz. 1.00			
" oxide, red, (tri-oxide; formerly called: sesqui-oxide), [Uranic Oxide; Uranyl Oxide; Anhydrous Uranic Acid], pure	oz. 1.50			
" phosphate	oz. 1.00			
" sulphate	oz. .85			
Uranium Yellow, see Sodium, uranate; *and also:* Ammonium, uranate				
Urari (Woorari, Woorara, Woorali), see Curare				
Urea (Carb-amide), pure, cryst.	oz. .75			
" acetate, fused	oz. 1.50			
" citrate	oz. 1.75			
" hydrochlorate	oz. 1.75			
" nitrate	oz. .75			
" oxalate	oz. .75			
" sulphate	oz. 1.75			
Urea, Acetylene-, see Acetylene-urea				
" **Sulpho-,** see Sulpho-urea				
Ur-ethane (Ethylic Urethane), chem. pure, Merck,—(*Carb-amate of Ethyl*)	oz. .60			
" **Ethylidene-,** chem. pure	oz. 2.00			
" **Chloral-,** chem. pure, cryst.	oz. 6.00			
Ur-ethylane (Methylic Urethane), chem. pure	oz. 2.00			
Uro-bilin (Hydro-bili-rubin[-phain])	1½ gr.vl. 10.00			
Uro-melanin,—according to Thudichum	1½ gr.vl. 10.00			
Urson, chem. pure	15 gr. 1.00			

☞ When ordering, specify: **"MERCK'S"** !

	Containers incl.
Vanadium, metallic, fused...............	15 gr. 22.00
" chloride...........................	⅛ oz.vls.oz. 3.00
". pent-oxide, hydrated, (Vanadic Hydroxide), see Acid, vanadic, Meta-........	
Vanillin, synthetic.—1 part, in alcoholic dilution or sugar-trituration, represents 40 parts of best Vanilla Bean................	oz. 6.50
Vaselin (Cosmolin), yellow,—melting-point 40-42° C [104-107.6 F].............	
" white,—m.-p. 43-45° C [109.4-113 F]...	
" —for veterinary purposes.............	
" —Pennsylvania.....................	
Vasicine.—Alkaloid from Adhatoda vasica, Nees.—(A bronchial remedy, and insecticide.)......................................	
Vellozin (Vellosin), see Vieirin	
Veratrine Merck, (Veratria):	
pure...................................	⅛ oz.vls.oz. 1.55
chem. pure,—conform. to *U. S. Ph.* and Ph. G. II................................	⅛ oz.vls.oz. 1.65
acetate................................	⅛ oz.vls.oz. 2.00
hydrochlorate	⅛ oz.vls.oz. 2.00
nitrate................................	⅛ oz.vls.oz. 1.75
sulphate..............................	⅛ oz.vls.oz. 1.75
valerianate	⅛ oz.vls.oz. 1.75
Verdigris, purified, see Copper, acetate, basic	
" crystallized, see Copper, acetate, normal, *U. S. Ph.*	
Verditer, blue, see Copper, carbonate, blue	
Vermilion, artificial, best, see Mercury, sulphide, red, *U. S. Ph.*...................	
Vernonin, — [$C_{10}H_{24}O_7$]. — Glucoside from the root of Vernonia nigritans, S. & M., (South-east African "Batjentjos");—deliquescent powder.—[Mild heart-tonic.]....	
Vesuvine, see under Aniline and Phenol Dyes: Brown............................	
Vieirin (Vieiric Acid) [Vellozin; Cuprein],—from the bark of Remijia Vellozii, De Candolle, (Cuprea-bark). — [A febrifuge highly valued in the Brazils.].............	15 gr. 3.00
Vienna Caustic, powder, see Potassium, hydroxide, with Lime, [2:1], powder.................	
" " fused, (Filhos's Caustic), see do., do., do. do., [4:1], fused	
Vinegar, concentrated, pure, (Acetum concentratum purum), see Acid, acetic, pure,—solution	
" do., chem. pure, (Acetum purissimum, Ph. G. II), see Acid, acetic, chem. pure, —solution.......................	
Vinegar, pyroligneous, (*Wood-vinegar*), rectified, [Acetum pyrolignosum rectificatum, Ph. G. II], see Acid, pyro-ligneous, purified...................................	
Vinegar of Lead, ("Goulard's Extract"), see Solutions: Lead acetate, basic, *U. S. Ph.*	
Vinegar Naphtha, see Ether, acetic......	
Vinum Opii, — so-called, — see Tinctures: Opium,—saffronated	
" **Pepsini,** Ph. G. II, see Pepsin Wine..	
Viride Æris purificatum, see Copper, acetate, basic	
Vitellus (Vitellus Ovi), see **Egg preparations:** Yelk, etc.....	
Vitriol, blue (*Copper-*), see Copper, sulphate, neutral, *U. S. Ph.;* and other grades and forms	

☞ **When ordering, specify: "MERCK'S"!**

	Containers incl.			
Vitriol—(*continued!*),—green (*Iron*-), see Iron, sulphate, Ferrous : *U. S. Ph.; do.* precipitated; *do.* exsiccated;—and other grades and forms				
" Lead-, see Lead, sulphate, etc.				
" white (*Zinc*-), see Zinc, sulphate, *U. S. Ph.*; and other grades and forms				
Vitriol, so-called "**Oil**" of ; free from Arsenic;—see Acid, sulphuric, crude				
Vitrum Antimonii (*Stibii*), [Antimonial Glass], see Antimony, sulphide, vitreous,—so-called				
" **Arsenii**, (Vitreous Arsenic ; Arsenic-glass), see Acid, arsenious,-pure, *lumps*				
" **Boracis**, (Vitrified Borax; Borax-glass), see Sodium, bi-borate, fused				
Volumetric Solutions, *pharmacop'l*, (Test-solutions), see at End of List.				
Vomicine, see Brucine				
Water (Aqua), Acorn,—acc. to Rademacher	lb.	.50		
" Almond, Bitter-,-(Aqua amygdalæ amaræ),—Ph. G. II	lb.	.40		
" Asafetida,—(Aqua Asæ fœtidæ),—simple	lb.	.75		
" Balm (*Lemon-balm*), see Water, Melissa				
" Cherry-laurel, see Water, Laurel, Cherry-				
" Chlorine, see Chlorine-water				
" Cinnamon; alcoholized,—(Aqua Cinnamomi spirituosa [vinosa])	lb.	.50		
" fetid antihysteric, compound, — (Aqua fœtida anti-hysterica composita, Ph. G. I)	lb.	1.00		
" hydrosulphuretted	lb.	.50		
" Laurel, Cherry-, — (Aqua Laurocerasi, Ph. G. I)	lb.	.40		
" Lime, see Solutions: Lime, *U. S. Ph.* .				
" Melissa (Balm, Lemon-balm),—decuple	lb.	1.00		
" Opium; highly concentrated,-quintuple	lb.	1.25		
" Quassia,—acc. to Rademacher	lb.	.50		
" Tobacco,—(Aqua Nicotianæ),—acc. to Rademacher	lb.	.60		
" Vomic-nut,—(Aqua Nucum vomicarum),—according to Rademacher	lb.	.50		
Water of **Ammonia**, see Ammon., Water of				
Water, oxygenated, — so-called, — see Hydrogen Per-oxide, etc.; etc.				
Water-glass (*Soluble* Glass and *Liquid* Glass), see Potassium, silicate, etc.; and, Sodium, silicate, *U. S. Ph.*; etc.				

☞ **When ordering, specify : "MERCK'S"!**

	Containers incl.		
Wax Paper, see under Paper............			
Waxed Sponge, see Sponge-tent..........			
Whey, so-called "**Essence**" of, see Rennet Wine...................................			
Wine of Opium, — so-called, — see Tinctures: Opium; saffronated..........			
" of **Pepsin**, Ph. G. II, see Pepsin Wine			
" of **Rennet**, see Rennet Wine.........			
Wolfram (Wolframium, Tungsten), metallic, chem. pure.......................	15 gr. .30		
" metallic, commercial.................	lb. 1.50		
" oxide, tri-, (Wolframic [Tungstic] Oxide), see Acid, wolframic, anhydrous.			
Wood-oil, so-called, ("East-Indian Wood-oil," or: "East-India Copaiva Balsam," so-called), see Balsams: Gurjun............			
Wood-spirit (Wood-naphtha, Wood-alcohol), see Alcohol, methylic..............			
Wood-vinegar, rectified, see Acid, pyroligneous, purified.......................			
Wool, Philosophers', — so-called, — see Zinc, oxide, by dry process.............			
Woorali (Woorara, Woorari), see Curare....			
Xanthine (Xanthin), [Xanthic Oxide; Ureous Acid, Uric Oxide].....................	15 gr. 10.00		
Xylene (Xylol), [Di-methyl-benzene], pure, —b.-pt. 137–140° C [278.6–284 F].........	lb. .85		
Xylidine (Amido-xylene [-xylol])...........	oz. .30		
Xylostein................................	1½ gr. vial 2.00		
Yelk (*Yolk*) [Vitellus], of egg,—dried,—see under **Egg preparations**....................			
Yttrium, metallic........................	15 gr. 9.00		
" carbonate.....................	15 gr. 2.00		
Yttrium and Platinum, cyanide, see under Platinum double Cyanides			

☞ When ordering, specify: "**MERCK'S**"!

		Containers incl.			
Zinc (Zincum), *Amalgams* and *alloy* of, see *after the double salts*,—[below!]					
·· double salts of, see "Zinc and-" (below!)					
·· metallic, absolutely chemically pure...	lb.	3.00			
·· " highly pure, granulated	lb.	1.60			
·· " " " in sticks	lb.	1.60			
·· " " " powder	lb.	1.75			
·· " absolutely free fr. Arsenic,—granulated;—*Zincum, U. S. Ph.*	lb.	.50			
·· " absol. free fr. Arsenic,—in sticks..	lb.	.55			
·· " " " " —coarse powd.	lb.	1.00			
·· " powder, (Zinc-dust)	lb.	.30			
·· " blocks,—for Hydrogen lamps...	lb.	.40			
·· " crude, in sticks	lb.	.40			
·· acetate, pure,—*U. S. Ph.* and Ph. G. II	lb.	.57			
·· " " fused	lb.	.50			
·· albuminate	oz.	.50			
·· arseniate (arsenate)	oz.	.30			
·· arsenite	oz.	.25			
·· benzoate,—from *true* Benzoic Acid, prepared from the resin	oz.	.59			
·· " —from *artificial* Benzoic Acid....	oz.	.40			
·· bi-borate	oz.	.30			
·· borate	oz.	.25			
·· bromate	oz.	1.00			
·· bromide,—*U. S. Ph.*	oz.	.23			
·· carbonate, precipitated,—*U. S. Ph.*	lb.	.50			
·· chlorate	oz.	.50			
·· chloride (muriate), [Butter of Zinc], fused, in sticks;—*U. S. Ph.*	oz.	.13			
·· " fused, in troches	oz.	.15			
·· " dry, white,—*U.S. Ph.* and Ph. G. II	oz.	.13			
·· " crude, dry	lb.	.30			
·· " " liquid,—aqueous solution..	lb.	.30			
·· " " " —alcoholic solution..	lb.	.50			
·· " fused, with Potassium Nitrate....	lb.	1.50			
·· chloro-iodide	oz.	.75			
·· chromate	oz.	.30			
·· citrate	oz.	.40			
·· cyanide...) ("Zincum cyanatum *sine*	oz.	.27			
·· " pure) *Ferro*")	oz.	.50			
·· ferro-cyanide, (Zincum zoöticum [borussicum]), ["Zincum cyanatum *cum Ferro*"]	oz.	.27			
·· gynocardate.—(Dermatological remedy.)	½ oz.vl ..oz.	2.00			
·· hypo-phosphite	oz.	.70			
·· **ichthyol-sulphonate**, see under **Ichthyol** prep.					
·· iodate	oz.	1.50			
·· iodide,—*U. S. Ph.*	oz.	.52			
·· lactate	oz.	.34			
·· mono-chlor-acetate, cryst	15 gr.	.50			
·· muriate, see Zinc, chloride, *U. S. Ph.s;* and other grades and forms					
·· nitrate, crude	lb.	.75			
·· " pure	oz.	.25			
·· oleate	oz.	.35			
·· oxalate	lb.	1.00			
·· oxide, by wet proc., white, chem. pure.	lb.	.70			
·· " " " " " —*U.S.Ph.* and Ph. G. II...	lb.	.65			
·· " " " " " II	lb.	.60			
·· " by dry process, (Flowers of Zinc; so-called "Philosophers' Wool"; Nihil album)	lb.	.25			
·• **per-manganate, liquid,—[25%]**	oz.	.40			
·· " chem. pure, cryst,—a highly pure, well crystallized preparation;—free fr. Potassium Per-mangan., Chlorine, Sulphuric Acid, etc...	oz.	.94			

☞ When ordering, specify: "**MERCK'S**"!

	Containers incl.			
Zinc, phosphate, cryst.	oz. .18			
" phosphide (phosphuret), lumps } U. S.	oz. .77			
" " powder } Ph.	oz. .77			
" phosphite	oz. .65			
" picrate (picro-nitrate)	oz. .35			
" pyro-phosphate	oz. .30			
" salicylate, white	oz. .49			
" silicate	oz. .45			
" sulphate, (Zinc Vitriol; White Vitriol), pure, cryst.,—U. S. Ph.	lb. .31			
" " pure, dry	lb. 1.00			
" " in sticks.	oz. .40			
" sulphide (sulphuret), pure	oz. .30			
" " commercial	lb. .75			
" sulpho-ichthyolate, see under Ichthyol preparations				
" sulpho-phenate (phenol-sulphonate, sulpho-carbolate), cryst.,—[Para-phenol-sulphonate of Zinc],--Ph. G. II	oz. .14			
" tannate	oz. .30			
" tartrate	oz. .40			
" tri-chlor-phenate	oz. .75			
" valerianate, cryst., light,—U. S. Ph.	oz. .35			
" " powder	oz. .30			
Zinc and Aluminium, sulphate, see Alum, zincic				
" and Ammonium, chloride	oz. .60			
" and Iron, cyanide, so-called, see Zinc, ferro-cyanide				
" and Manganese, chloride	lb. .75			
" and Mercury .... } Amalgams.—see Zinc Amalgam; and, Zinc and Tin,				
" " " and Tin, } Amalgam;—(below!)				
" and Potassium, cyanide, cryst.	oz. 1.00			
Zinc Alum, see Alum, zincic				
" Amalgam	lb. 1.50			
" and Tin, Amalgam	lb. 2.00			
" -Sodium alloy	oz. .50			
" Vitriol, (White Vitriol), see Zinc, sulphate, U. S. Ph.; and other grades and forms				
Zinc, Butter of, see Zinc, chloride, U. S. Ph.s; and other grades and forms				
" Dust of, see Zinc, metallic, powder				
" Flowers of, see Zinc, oxide, by dry process				
Zirconium, metallic, cryst.,—fine leaflets	15 gr. 10.00			
" oxide	15 gr. 1.10			
" sulphate	15 gr. 1.00			
Zirconium and Potassium, fluoride	15 gr. .50			
Zymase, see Invertin				

N.B.—See next page for "Specimen Collections" and "Test-Solutions";—page 155 for "Merck's Guaranteed Reagents";—and page 156 for Table of Abbreviations.

☞ When ordering, specify: "MERCK'S"!

SPECIMEN COLLECTIONS.

	Containers incl.			
Alkaloids—(52 Specimens): —in tubes of 1-gramme liquid capacity —" " " ½ " " " **Alkaloids, Glucosides, etc.**—(72 Specimens): —in tubes of 1-gramme liquid capacity —" " " ½ " " " **The Opium constituents, complete,**—*embracing 23 Alkaloids, etc.,* in QUANTITIES CORRESPONDING to the average proportions in which they NATURALLY OCCUR in the Crude Drug.................. **Metals**—(61 Specimens) .... **Physiological Preparations**—(42 Specimens)	38.00 20.00 45.00 23.50 20.00 20.00			

In elegant Cases.

TEST-SOLUTIONS.

for Qualitative and Quantitative Analyses.

Indicator Solutions:
  Chameleon Mineral, (*Manganate of Potassium*).—Titration *not* guaranteed ......
  Cochineal, — hydro-alcoholic, [3 : 250],— Ph. G. II..............
  Lacmus (*Chemically Pure Litmus*), for alkalimetry,—titrated ....
  Phenol-phtalein, — alcoholic, [1 : 100],— Ph. G. II.............

Titrated Normal Solutions, for quantit. analyses:
  Acid, nitric,—normal, $=\frac{1}{1000}$ equivalent of alkaline earth..................
  " oxalic,—normal, $=\frac{1}{1000}$ equivalent of alkali....................
  " sulphuric, — normal, $=\frac{1}{1000}$ equivalent of alkali..................
  Arsenic, — (Arsenious Oxide, Anhydrous Arsenious Acid), — deci-normal, $=\frac{1}{10,000}$ equivalent of Chlorine ....
  Barium Chloride,—normal..............
  Copper Tartrate, potassic, — (*Fehling's* Solution) ....
  Iodine........................
  Mercuric Nitrate,—1 cub. cm. = 0.01 gramme Urea........................
  Potassa, caustic, — normal, $=\frac{1}{1000}$ equivalent of acid ....
  Silver Nitrate, — deci-normal, $=\frac{1}{10,000}$ equivalent of Bromine or Chlorine ......
  Soap,—acc. to Clark.—Titration *not* guaranteed........................
  Soda, caustic,—duplo-normal,—for Vinegar tests ....
  Sodium Chloride,—deci-normal, $=\frac{1}{10,000}$ equivalent of Silver ....
  Sodium Thio-sulphate ("*Hypo-sulphite*"),— deci-normal....................
  Uranic Acetate,—1 cub. cm.=0.005 gramme P_2O_5....................
  Uranic Nitrate,—1 cub. cm.=0.005 gramme P_2O_5....................

Pharmacopeial Volumetric Solutions, — according to *U. S. Ph.* or to Ph. G. II., etc.,—furnished to order.

☞ **When ordering, specify: "MERCK'S"!**

MERCK'S GUARANTEED REAGENTS.

N.B.—These Reagents are supplied by me under STRICT GUARANTEE of their ABSOLUTE CONFORMITY to the STANDARDS OF PURITY established by DR. C. KRAUCH'S TREATISE on "PURITY-TESTS FOR CHEMICAL REAGENTS."—In order to obtain them under the GUARANTEE stated, it will be necessary to SPECIFY, in each instance:—"MERCK'S GUARANTEED REAGENTS."

Acid, acetic, ch. p., conc., [1.064]
" carminic, pure
" chromic, ch. p.; *free fr. Sulphuric Acid*
" citric, perfectly white, ch. p., cryst.
" hydrochloric, pure, [1.19]
" hydrofluoric, fuming, ch. p.
" hydro-silico-fluoric, ch. p.
" molybdic, pure
"   " ch. p.; *free fr. Ammonia*
" nitric, pure, [1.20]
"   " fuming, pure, [1.48]
" oxalic, ch. p.
" phospho-molybdic,—solution
"   " -wolframic (*-tungstic*),—solution
" pyro-gallic, re-sublimed
" sulphuric, ch. p., [1.84]
"   " fuming
" tannic, see *Tannin*
" tartaric, ch. p., cryst.
Alcohol, absolute, pure, [0.796]
" amylic, ch. p.
" methylic, ch. p.
Ammonia, Water of, pure, [0.925],—abt. 20%
Ammonio-*Ferrous* Sulphate
Ammonium, carbonate, ch. p.
" chloride, pure
" fluoride, ch. p.
" molybdate, ch. p.
" nitrate, ch. p.
" oxalate, ch. p.
" sulphate, ch. p.
Aniline, pure
Barium, acetate, ch. p.
" carbonate, ch. p.
" chloride, ch. p.
" hydroxide ("hydrate"), [Caustic *Baryta*], ch. p., cryst.
" nitrate, ch. p.
Bismuth, hydroxide (hydrated tri-oxide), pure
Calcium, chloride, ch. p., cryst.
"   " pure, dry
"   " oxide, caustic, (Burnt *Lime*),—*from marble*
"   "   "   —*from Iceland spar*
" sulphate, pure, precipitated
Carbon Bi-sulphide, ("*Alcohol Sulphuris*"), pure
Chloroform, pure
Cobalt, nitrate, ch. p.
Copper, metallic, ch. p.
" oxide (mon-oxide), pure, powder
"   "   "   " coarse granules
" sulphate, ch. p., cryst.
Di-phenyl-amine, ch. p.
Ether, ch. p., [0.720–0.722]
"   "   " anhydrous; *distilled over Sodium*
Hydroxyl-amine, hydrochlorate, ch. p.
Iodine, re-sublimed, ch. p.
Iron, chloride, *Ferric*, (sesqui-[tri-]chloride)
" sulphate, *Ferrous*, ch. p., cryst.
" sulphide (sulphuret), *Ferrous*,—lumps
"   "   "   "   —sticks
Iron *and* Ammonium, sulphate,—*Ferrous*,—see Ammonio-Ferrous Sulphate
Lead, acetate, ch. p.
" chromate, pure
" oxide, *yellow*, (mon-oxide), [*Litharge*], ch. p.
Magnesium, carbonate
" chloride, ch. p.
" oxide, (Calcined *Magnesia*)
"   " *free fr. Sulphuric Acid*
" sulphate, ch. p.

Manganese, per-oxide, *native*, [Black Oxide], [Pyrolusite].—lumps
Mercury, bi-chloride, (Corr. Sublimate), ch. p.
" nitrate, *Mercurous*, ch. p.
" oxide, *Mercuric*, yellow (by *wet* process), [*Yellow* Precipitate], ch. p.
Paper, Litmus- ; red *or* blue
Platinum, tetra-chloride (per-chloride), [Platinic Chloride],—*formerly* called *bi*- or *di*-chloride; - dry, pure
Potassium, antimonate, pure
" bi-chromate, ch. p., cryst.
" bi-sulphate, ch. p., cryst.
" bromate, ch. p.
" carbonate, ch. p.
" chlorate, ch. p.
" chromate, *yellow*, ch. p.
" cyanide, ch. p.
" *ferrid*-cyanide, (*Red* Prussiate of Potassa)
" *ferro*- "   (*Yellow* "   "   ")
" hydroxide ("hydrate"), [Caustic *Potassa*], ch. p.
" do., pure (*purif. by Alc.*),—sticks *or* lumps
"   " purified,—sticks *or* lumps
" iodide, ch. p.
" nitrate, ch. p.
" nitrite, ch. p.
" per-manganate, pure, cryst.
"   " ch. p.; *free fr. Sulphuric Acid*
" sulphate, ch. p.
" sulpho-cyanate (*thio-cyanate*; *rhodanide*), ch. p.
Silver, metallic, ch. p., - sheet
" nitrate, ch. p.,—cryst. *or* sticks
Sodium, acetate, ch. p.
" bi-borate, pure, cryst., *prismatic*, (Officinal Refined *Borax*)
" bi-carbonate, ch. p., powder
" bi-sulphate, ch. p., cryst.
" bi-sulphite, pure, dry
" carbonate, ch. p., cryst.
"   "   "   " dry
" chloride, ch. p.
" hydroxide ("hydrate"), [Caustic *Soda*], ch. p.,—*from Sodium*
" do., pure (*purif. by Alc.*),—sticks *or* lumps
"   " purified,—sticks *or* lumps
" nitrate, ch. p.
" nitrite, ch. p.
" thio-sulphate 'so-c. "*hypo-sulphite*"), ch. p.
" wolframate (*tungstate*), ch. p.
Sodium *and* Ammonium, phosphate, pure
Solution of Ammonia, aqu., see Amm., Water of
" of Ammonium Sulphide, hydrosulphuretted,—(*Hydrothion-Ammonium* solution)
" of Indigo Sulphate
" of Potassium Hydroxide, pure, [1.30]
" of Sodium Hydroxide, crude, [1.30]; *free fr. Nitrogen*
"   " do. do., pure, [1.30]; *free fr. Nitrogen*
Tannin (*Tannic Acid*), ch. p.
Tin, chloride, (*true bi*-chloride), pure, cryst.
Uranium, nitrate, ch. p.
Water of Ammonia, see Ammonia, Water of
Zinc, metallic, ch. p.,—granulated *or* sticks
"   "   " —powder
"   "   " —*absolutely free fr. Arsenic*,—sticks
"   "   " —*do. do. do. do.*,—granulated
"   "   " — "   "   "   " —coarse powder
"   "   " powder, (Zinc-dust)

☞ For TABLE OF ABBREVIATIONS,—*see next page!* ☜

ABBREVIATIONS
OCCASIONALLY EMPLOYED IN THE PRECEDING LISTS.

The abbreviation:	Means:
ab. *or* abt.	about
abs.	absolute
Ac.	Acid
acc.	according
Alc.	Alcohol
alc. *or* alco.	alcoholic
anh. *or* anhyd.	anhydrous
Aq. *or* aq.	Aqua (Water, = H_2O)
aqu. *or* aque.	aqueous
artif.	artificial
°B *or* °Bé.	degrees of Baumé's hydrometer
bot's	bottles
b.-p. *or* boil.-pt.	boiling-point
°C	degrees of Celsius's (*centigrade*) thermometer
cbcm *or* cub. cm	cubic centimetre[s] (= 16.2316—or, about 16¼—minims)
cg	centigramme[s] ($^1/_{100}$ of a gramme) [= 0.1543—or, about $^{15}/_{100}$—of a grain]
ch. p. *or* ch. pure	chemically pure
cm	centimetre[s] (= 0.3937—or, about $^4/_{10}$—of an inch) !
com'l *or* comm'l	commercial
comp. *or* comp'd	compound
conc.	concentratus (*or* concentrated)
conf.	conforming
cont.	containing
contin.	*continued*
corr.	corrosive
depur.	depuratus (= purified)
diss.	dissolves
div. spec.	divers species
eff. *or* efferv.	effervescent (effervescing)
emp. *or* empyr.	empyreumatic
eth. *or* ether. *or* eth'l	ethereal
Ex. *or* Ext.	Extract
expr.	expressed
F—(*degree-mark omitted!*)—	degrees of Fahrenheit's thermometer
Fl. Ex. *or* Fl. Ext.	Fluid Extract
fr.	from
gm	gramme[s] (= 15.4325—or, about 15½—grains)
gr.	grain (*or* grains)
gran.	granulated *or* granules
hyd.-alc. *or* hydro-alco.	hydro-alcoholic
ident.	identical
imp. pwd.	impalpable powder
insp.	inspissated
lge.	large
Lic.-r. *or* Licor.-rt.	Licorice-root
Liq.	Liquor (= Solution)
liq.	liquid
mg	milligramme[s] ($^1/_{1000}$ of a gramme) [= abt. $^1/_{65}$ of a grain]
mm	millimetre[s] (= 0.039—or, about $^4/_{100}$—of an inch)
mol. *or* molec.	molecule (*or* molecules)
m.-p. *or* melt.-pt.	melting-point
mtd.	mounted
orig.	original
perf. *or* prf.	perfectly
Ph. Au. *or* Ph. Austr.	Pharmacopœia Austriaca, of 1869 ; and Additions of 1879
Ph. Belg.	" Belgica, of 1885
Ph. B. (*or* Ph. Bor.) V; (—VI)	" Borussica, of 1829 ; (—of 1846)
Ph. Br. *or* Ph. Brit.	" Britannica, of 1867
Ph. Br. n. *or* Ph. Brit. new	" " 1885
Ph. G. I	" Germanica, of 1872
Ph. G. II	" " 1882
Ph. Helv.	" Helvetica, of 1872 ; and Additions of 1876
Ph. Hung.	" Hungarica, of 1871
Ph. Nl. *or* Ph. Neer.	" Neerlandica, of 1871
Ph. Port.	" Portugallensis, of 1876
Ph. Ross.	" Rossica (*Russica*), of 1880
pharm. *or* pharm'l	pharmacopeial (pharmacopœial)
prec. *or* precip.	precipitated *or* precipitate
prep.	preparation[s] *or* prepared
prep'd	prepared
prf.	(*see* perf.)
proc.	process
purif.	purified
puriss.	purissimus (= chemically pure)
pwd.	powder *or* powdered
rect.	rectified
sm. *or* sm'l	small
so-c. *or* so-c'd	so-called
Sol. *or* sol.	Solution (*or* Solutions)
s.-p. *or* solid.-pt.	solidifying-point
sp. gr.	specific gravity
sym. *or* symm.	symmetrical
und.	under
U. S. Ph.	United-States Pharmacopœia, of 1882
U. S. Ph. of 1870	" " " 1870
U. S. Ph.s	a group of two or more *U.-S.-Ph.* preparations
vl. (vls.)	vial (vials)
W.	Water
w.	with
wh.	white

N.B.—*Besides these,* the names of various substances in the List, when repeated soon after their occurrence in full print, are sometimes abbreviated, where their meaning is evident; as, for instance,—on page 14:—*after* "Ammoniated Glycyrrhizin," the letters "Gl." occurring in the latter part of the line, of course, mean "Glycyrrhizin"; or, as,—on page 16:—*after* "Ammonium and Cobalt, sulphate," the abbreviation "C. & A., sulph." will be readily understood as meaning: "Cobalt and Ammonium, sulphate."

PURE DRUGS,

Are always obtainable, and have been for nearly a hundred years past,

OF

W. H. Schieffelin & Co.,

170 & 172 William Street,

NEW YORK.

Works at Feuerbach
Established by FRIDR. JOBST, 1806.

Works at Sachsenhausen
Established by C. ZIMMER, 1837.

Branch House at Milan,
"Successori di FRIDR. JOBST."

*Vereinigte Fabriken Chemisch - Pharmaceutischer Producte,
Feuerbach-Stuttgart u. Frankfurt a. M.,*

ZIMMER & CO.,

FRANKFORT o. M., Germany.

ZIMMER'S QUININE,

QUINIDINE, CINCHONIDINE, CINCHONINE.

THIS

SUPERIOR

Brand is Represented in the United States of America by

E. MERCK'S U.-S. HOUSE,

73 William Street, NEW YORK.

☞ __PRICES__ will compare favorably with those of any other reputed brand.

ESTABLISHED 1851.

EIMER & AMEND,

Nos. 205, 207, 209 and 211 Third Avenue,

NEW YORK.

18th Street Station of Elevated R. R.

MANUFACTURERS AND IMPORTERS OF

STRICTLY CHEMICALLY PURE CHEMICALS, ACIDS,

AND

CHEMICAL APPARATUS.

Only uptown house carrying FULL LINE OF MERCK'S GOODS on hand.

Polariscopes, Hammered Platina, Nickel Ware.	Assay Goods, Bunsen's Burners, Combustion Furnaces, Agate Mortars, Copper Stills, etc.
Glass Blowing and Engraving done on premises.	

We carry the heaviest and best selected stock of Chemical Apparatus, Platina Goods, Filter Papers, Bohemian Glass, Royal Berlin China, Acid Proof Stoneware, Balances and Weights in the U. S. Being the sole representatives of the following large and world renowned Manufacturers, our facilities are unlimited.

C. Schleicher & Schüll's German Filter Papers.

E. March Söhne, German Chemical Stoneware.

Joseph Kavalier's, Infusible Bohemian Glass.

LeBrun, F. Desmoutis & Co.'s Chem. pure Hammered Platina.

H. Fleitmann's, Wrought Nickel Ware.

Greiner & Friedrich's, German Glass Ware.

G. Kern & Sohn, German Balances and Weights.

Dr. C. Scheibler's, Standard Sugar-Testing Instrument.

THEODORE METCALF & CO.,
Pharmaceutical Chemists,
39 Tremont Street,   BOSTON, MASS.

OLEUM LANAE METCALF.
PURE ODORLESS WOOL FAT.

Free from WATER, ANIMAL and VEGETABLE FATS.

NEVER MOULDS OR SEPARATES.

Less our usual Trade Discount.

One-lb. Cans, 1 dozen in case,	50 cts per lb.
Five-lb. Cans, ½ dozen in case,	45 " "
Fifty or One Hundred-lb. cans, boxed,	30 " "

We pay Freight ONLY on lots of Fifty Pounds.

FOR SALE BY ALL WHOLESALE DRUGGISTS.

We solicit correspondence for the various grades, as we can furnish any color or melting-point.

METCALF'S COCA WINE FROM FRESH COCA LEAVES.

The superior quality of wine and coca has commended this wine to the physicians and buyers of the United States.

............*Send for REBATE circular and price List.*............

ALKALOIDS, CHEMICALS, METALS,
FINE, RARE and CRUDE of every description.

From the many years we have dealt in this class of supplies, we claim to be leaders in this branch of the drug trade, and by constantly replenishing and increasing our stock, and at once procuring or manufacturing all new chemical products, we are able to do full justice to all orders or requests for quotations.

We make a specialty of the products from the Laboratories of

T. MORSON & SON,	**LONDON.**
E. MERCK,	**DARMSTADT.**

Peninsular White Lead and Color Works,

DETROIT, MICH.

DRY COLOR MAKERS and Manufacturers of FINE PURE PAINTS

For House Painting and Decorating.

SUPERFINE COACH COLORS IN JAPAN.
FINE PURE COLORS IN OIL.
RAILWAY PAINTS,
PRIMING, FILLING, ROUGH STUFF AND SURFACERS.

PURE READY MIXED PAINTS.
WHITE AND TINTED LEADS.

CARRIAGE AND BUGGY PAINTS,
Ready for Use, Quick Drying, in Nine Choice Colors.

DIPPING AND PASTE PAINTS,
For Wagon Makers and Agricultural Implements Manufacturers.

We offer special inducements to large buyers on Peninsular Permanent Red, Vermilions, Chrome Greens and Yellows, Maple Leaf Permanent Green, Prussian, Chinese, Steel and Soluble Blues, Rose Pink, Lakes, Pulp Colors, Wall Paper Colors and other specialties for Grinders, Paper Manufacturers, Lithographers, etc.

Peninsular Non-Corrosive Iron Filler and Steel Color Paints (four shades) for Founders, Machinists, Engine Builders, etc., and Peninsular Wood Fillers are warranted to give satisfaction in every respect.

CORRESPONDENCE SOLICITED.

FARRAND, WILLIAMS & CO., General Agents,

DETROIT, MICH.

Factory: Lieb Street, from Transit R. R. to River Front.

ESTABLISHED 1815.

FARRAND, WILLIAMS & Co.,
IMPORTING AND MANUFACTURING WHOLESALE

DRUGGISTS
AND DEALERS IN
DRUGGISTS' SUNDRIES.

State Agency and Depot for all the Leading Patent Medicines.

Our own Importations of Crude Drugs, Essential Oils, Olive Oil, Chamois Skins, Hair, Tooth and Nail Brushes, Etc.

Are offered to the Drug Trade in competition with Eastern markets. We grind and powder our own Drugs from choicest selections, and can therefore guarantee their quality, as well as the reliability of our Fluid Extracts, Elixirs, Medicated Syrups and Fine Pharmaceutical Preparations.

Orders by Mail a Specialty. We protect the retail trade by not selling to consumers, we fill all orders promptly and completely for all goods in our line, and obtain other goods if they are to be had in our market.

We carry the greatest variety and the largest stock of any house in the State in our line. The fitting up of new drug stores complete a specialty.

FARRAND, WILLIAMS & CO.

ANTIFEBRIN

(KALLE'S)

AS A

1. **HYPNOTIC**, ANODYNE, SEDATIVE.
2. TONICO-**NERVINE**.
3. ANTI-**EPILEPTIC**.
4. MITIGANT OF **VARIOLA** VERA.
5. ANTI-ARTHRITO-**RHEUMATIC**.

According to most distinguished medical testimony, ANTIFEBRIN, in the above and many other applications, has SUPERSEDED THE FOLLOWING older remedies in Efficacy or in Safety: *Quinine, Antipyrine, Potassium Bromide*, the *Iodides, Chloral Hydrate, Aconite, Morphine, Caffeine, Kairine, Salicylic Acid, Water*.

Beside the above, *Antifebrin continues to enjoy the decided preference of the Medical Profession* **OVER ALL OTHER ANTIPYRETICS**; being, f. i., of *Four Times the Strength* of Antipyrine,—according to the *Clinical Report* of Drs. A. Cahn and P. Hepp, of Prof. Kussmaul's Clinique at the University of Strassburg.

IODOLE

[Tetr-Iod-Pyrrole.]

—Containing over Eighty-eight Per Cent. of Iodine.—

SUCCEDANEUM —in All its External Uses.
— FOR — **IODOFORM** —Equaling it in Antiseptic Power.
—Preferable, being Entirely Pleasant and Safe !!!

IODOLE is { Wholly Odorless and Absolutely Non-Toxic. } ||| IODOFORM { has a Nauseating Odor and Poisonous Effects.

SOLE LICENSEE FOR THE U. S.: **E. MERCK**, NEW YORK.

THE TORSION BALANCE.

NO KNIFE EDGES.

NO FRICTION. NO WEAR.

Style 281.
Capacity 8 ounces on each pan. Sensible to 1/100th grain.
Six years' constant use of the Torsion Balance have proved it to be far superior to any form of Knife-edge balance.

Durable.
Accurate.

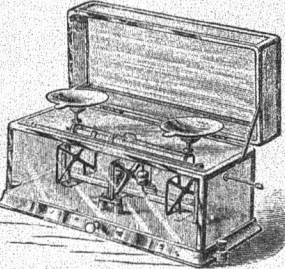

Sensitive.
Convenient.

Style 269.
Prescription Scale 3 inch german silver pans. Capacity 8 ounces, sensitive to 1/64th grain with rider beam graduated on upper edge from 1/8 grain to 8 grains, and on lower edge from 1/2 centigram to 5 decigrams.

Style 254.
Counter Scale. 9 inch pans.

Style 270.
Prescription Scale.

Write for Price List to

THE SPRINGER TORSION BALANCE CO.,
92 Reade Street, NEW YORK.

ESTABLISHED IN 1840.

HENRY TROEMNER,
710 Market Street,   PHILADELPHIA.

MAKER OF

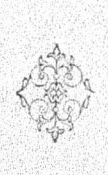

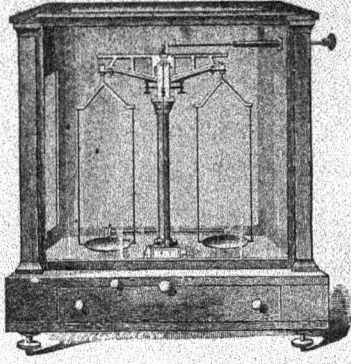

ANALYTICAL Balances, ASSAY Balances, &c., &c.

◁PRICE LIST ON APPLICATION.▷

Every Copper bears our Label and Seal "J.C."—guaranteeing the purchaser the finest Oil in every respect.

JAMES CHASKEL & CO.
DISTILLERS & IMPORTERS OF
ESSENTIAL OILS, ETHERS, DRUGS,
ESSENCES, CHEMICALS, ETC.

93 JOHN STREET, NEW YORK.

HAND PRESSED

OIL OF LEMON, J. C.

MANUFACTURED AT

MESSINA,

EXPRESSLY FOR

JAMES CHASKEL & CO.

OUR

16 Page

PRICE LIST

CONTAINING OUR

MANY

Specialties

MAILED ON APPLICATION

Singer & Wheeler,

IMPORTERS AND

Wholesale ✻ Druggists,

AND DEALERS IN

Druggist Sundries.

SALESROOMS: 420 TO 426 SOUTH WASHINGTON STREET,

WAREROOMS: 423 TO 429 SOUTH WATER STREET,

PEORIA, ILL.

Catalogues and special quotations furnished on application.

CREOLIN
— A SUCCEDANEUM —

For Carbolic Acid, Corrosive Sublimate, and for most of the

—GENERALLY PROPHYLACTICAL, SURGICAL, AND GYNECOLOGICAL—

DISINFECTANTS,

ANTISEPTICS, ANTI-ZYMOTICS,
BACTERICIDES, DEODORIZERS,
ASEPTICS, AND DETERGENTS.

CREOLIN

Is **NOT** a *Synthetic Compound*, but a *Natural Group*, of a number of the HIGHER PHENOLS, —including principally the CRESOLS,—abstracted from Coal-Tar Creasote.
Is **NOT** *Poisonous* (as Carbolic Acid and Corrosive Sublimate are), while *just as reliable and efficient* a Disinfectant and Antiseptic (in proper proportion); and a FAR MORE ENERGETIC Deodorizer than Carbolic Acid.
Is *readily miscible* — by Emulsion — with Water, and *very convenient* of application as a Wound-, Ulcer-, and Gangrene-Dressing.
In *combination* with MOLLIN, it forms *the best Aseptic Gynecological Lubricant*.
In VETERINARY Practice, it is *preferred* to all other Means of *Surface Disinfection*.

SOLE LICENSEE FOR THE U. S.: **E. MERCK**, NEW YORK.

ESTABLISHED 1856.

JULIUS ZELLER,

37 BOWERY,

P. O. Box 2824.   NEW YORK.

Importer, Exporter and Jobber

OF

FOREIGN & DOMESTIC DRUGS.

Pharmaceutical Preparations,
    Fine Chemicals and Rare Alkaloids,
        New Remedies and Essential Oils,
Prime Norwegian Cod-Liver Oil,
    True Dalmatian Insect Powder,
        Select Botanical Goods,

SOLID, FLUID AND POWDERED EXTRACTS.

FULLEST AND COMPLETEST STOCK CONSTANTLY ON HAND,

—— ALSO ——

OF

E. MERCK'S FINE CHEMICALS.

What is "Merck's Pepsin"?

—ANSWER:—

All those Various Grades and Forms of Pepsin

WHICH ARE REQUIRED BY THE DIFFERING

Pharmacopœias and Pharmaceutic Usages of All Nations!

This old fact has lately been utilized by Competitors for the purpose of creating confusion in the Public Mind on the above question; one of the *Lower Grades* made by MERCK *according to the official requirements of the Pharmacopœias,* being used by them in Competitive Tests, and falsely quoted by them as representing "MERCK'S PEPSIN" *generally.*—The above answer is therefore made, in order to correct those false representations.

None · of · the · Users · of · Any · of · MERCK'S · PEPSINS · have · ever had · occasion · to · ask · the · above · question; · for · they · know · what · they are · using; · and · this · is · precisely · The · Reason · Why · they · use · it.

THE PRINCIPAL VARIETIES
—OF—
MERCK'S PEPSIN

WILL BE FOUND ENUMERATED

In Page 106 of the Preceding List.

Specify "Merck's" Pepsins

FOR

EXCELLENCE, PERMANENCE, AND ECONOMY.

ESTABLISHED 1866.   INCORPORATED 1888.

HENRY HEIL CHEMICAL COMPANY,
ST. LOUIS, MO.
WHOLESALE ❋ DRUGGISTS,
IMPORTERS AND MANUFACTURERS OF
Chemicals and Chemical Apparatus.

SOLE AGENTS FOR:
J. H. Munktell's Swedish Filtering-Paper, Josef Kavalier's Unexcelled Bohemian Glassware, Royal Berlin and Berlin Porcelain Laboratory Ware, Battersea Crucibles, Scorifiers, Muffles and Furnaces, and L. Reimann's Superior Metal Goods.

AGENTS FOR:
Troemner's and Becker's Balances and Weights, Joseph Dixon's Blacklead Crucibles, Gundlach's Celebrated Hessian Crucibles, Freiberg Scorifiers, etc.

LARGE STOCK OF:
Galvanic Batteries, Insulated Wire, Gas-, Coal Oil- and Gasoline Laboratory Stoves, Filtering-Paper, Filter Pumps, Geissler's Air-Pumps for Electric Light Companies, Drying Apparatus, Platinum Wire, Foil and Ware, Hydrometers, Thermometers, Mortars of Steel, Iron, Brass, Porcelain, Wedgewood, Glass and Agate, Ringstands, Bunsen Burners, Blastlamps, etc.

WE CARRY THE MOST COMPLETE STOCK OF:
E. MERCK'S CELEBRATED CHEMICALS
FOR SCIENTIFIC, PHARMACEUTICAL, PHOTOGRAPHIC, MANUFACTURING AND TECHNICAL PURPOSES.

As we always have **ON HAND** all and every appliance and apparatus for chemical and metallurgical researches, we can supply **MORE PROMPTLY** than other houses all apparatus, utensils and materials used by

Smelters, Iron and Steel Manufacturers, Chemists, Assayers, Druggists, Miners, Manufacturing Jewelers, etc.

Commercial and Chemically Pure Acids a Specialty.

WE HANDLE ALSO LARGELY ALL HEAVY CHEMICALS, SUCH AS:
Copperas, Blue Stone, Plaster Paris, Blacklead, Cryolite, Silica, Tripoli, Tripoli Composition, Crocus, Argols, Alum, Saltpeter, Borax, White Lead, Bicarbonate of Soda, Hyposulphite of Soda, Soda Ash, Pearlash, Glauber Salt, Epsom Salt, Rochelle Salt, Carbonate of Ammonia, Sal Ammoniac, Tartaric Acid, Citric Acid, Oxalic Acid, Cream of Tartar, etc.

Before placing your orders, get our quotations!

HENRY HEIL CHEMICAL COMPANY,
212 SOUTH FOURTH ST.,   ST. LOUIS, MO.

JOHN WYETH & BROTHER,

MANUFACTURERS OF

Elegant * Pharmaceutical * Preparations,

EMBRACING

Medicinal Elixirs, Wines, Syrups, Liquors, Saccharated Pepsin, Pure Pepsin (Pepsin Porci), Absorbent Cotton, Suppositories, Medicinal Fluid Extracts,

COMPRESSED PILLS (OR POWDERS), COMPRESSED HYPODERMIC TABLETS, COMPRESSED TABLET TRITURATES, COMPRESSED MEDICINAL LOZENGES.

WYETH'S LIQUID EXTRACT OF MALT,
WYETH'S BEEF, WINE AND IRON,
WYETH'S DIALYSED IRON.

The above list is a synopsis of the character of the products of our establishment, and to which we beg to call the attention of manufacturing chemists, wholesale and retail druggists, and physicians. We will at all times be pleased to furnish Price Lists, Formulæ Lists, and circular matter pertaining to any and all of our preparations.

We were the originators of very many of the most extensively prescribed pharmaceutical combinations now in use, and were the pioneers in what is now generally termed Elegant Pharmacy. We were the first to introduce to the trade and medical men Compressed Pills, Compressed Hypodermic Tablets, Compressed Lozenges, and lastly, Compressed Tablet Triturates. The phenomenal favor with which they have been received is the best evidence of their value, as they are rapidly superseding all other kinds of Pills. Their wonderful accuracy, beauty of finish, ready solubility, permanency and ease of administration render them not only invaluable to the profession, but one of the greatest, if not the greatest, achievements in pharmacy of the age.

We are, perhaps, among the largest Manufacturers of Pepsin in the world, and claim for it a potency and digestive power greater than that of any produced.

Our list of Fluid Extracts embraces all the officinal, as well as unofficinal, drugs that are possessed of any medicinal value. They are made by a process peculiarly our own, from the most carefully selected, fresh, crude material; their exhaustion is absolutely complete, so that every pound represents a pound of drug.

Our Liquid Extract of Malt we have every reason to believe, from comparative tests and the high enconiums we have received from all sources, to be a preparation in every way worthy of all we claim for it, containing as it does a larger amount of nutritious malt extract, with less alcoholic spirit, than any other made.

Elixirs, Wines, Syrups, Liquors, Preparations of Beef, as made by us, are so well and favorably known they hardly require special mention. We claim for them absolute accuracy, careful and scrupulous attention to detail, and are in every respect just as represented on our labels.

We believe our Dialysed Iron is now recognized as the standard preparation of this article, supplying, as we do, not only the home market, but very extensively that of Europe.

Butter of Cacao Suppositories, we have for many years been large producers, and flatter ourselves, as made by us, every characteristic and requirement is fulfilled. Our list is very complete, embracing almost every variety of formulæ for the rectum, urethra, vagina, ear and nose. We are always glad to make any special formulæ desired.

Multum in Parvo!

A VALUABLE ADDENDUM

To ALL Medical or Pharmaceutical Periodicals

— IS: —

"MERCK'S BULLETIN,"

— A MONTHLY —

Record of New Discoveries, Introductions, or Applications of MEDICINAL CHEMICALS.

Moved by Professional — not Business — Interest!

SUBSCRIPTION PRICE: ONE DOLLAR PER ANNUM.

"MERCK'S BULLETIN" *Saves Time* to the busy Practitioner or Dispenser, in giving Prompt Information of Interesting and Valuable *Additions to the Materia Medica*, in the MOST CONCISE FORM POSSIBLE.

"MERCK'S BULLETIN" makes an *Exclusive Specialty* of reporting those *Advances in Chemical Art* which are of importance to the Physician and to the Druggist, — giving them UNMIXED WITH, AND UNACCOMPANIED BY, ANY OTHER MATTER WHATEVER.

"MERCK'S BULLETIN" is a *Thoroughly Reliable* Source of *Impartial and Exact* Information, — being edited WITHOUT ANY BIAS AS TO THE ORIGIN, MAKERSHIP, OR SELLING-INTEREST of any Substance discussed by it.

"MERCK'S BULLETIN" contains *No Advertisements or Business Notices* — either open or disguised; No Editorial Discussions, Correspondence, *Nor Any Expressions of View or Opinion;* — it consists *Solely* of a CONSCIENTIOUS COMPILATION OF ACTUALLY ASCERTAINED FACTS on NEW Developments in the MATERIA MEDICA.

☞ *Send your Address to* **E. MERCK,** New York, *for Free Sample Copy!*

**ONTARIO
COLLEGE OF PHARMACY**
ST. E.
RONTO.

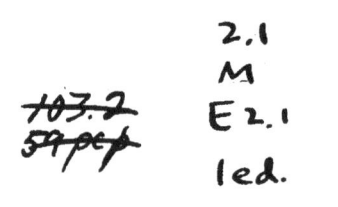

~~103.2~~
~~59 pcf~~

2.1
M
E 2.1
1 ed.

**PLEASE DO NOT REMOVE
CARDS OR SLIPS FROM THIS POCKET**

UNIVERSITY OF TORONTO LIBRARY

RS
51
M47
1889
c.1
PHAR

Lightning Source UK Ltd.
Milton Keynes UK
UKHW011858040219
336741UK00026B/1426/P